Information Resource Description

Information Resource Description

Editor

Silvius Mateus

scitus academics

Information Resource Description
Edited by **Silvius Mateus**

ISBN: 978-1-68117-557-7
Library of Congress Control Number: 2016942966

SCITUS Academics LLC,
www.scitusacademics.com
Box No. 4766, 616 Corporate Way,
Suite 2, Valley Cottage,
NY 10989

Printed in United States of America on Acid Free Paper

Preface

Information is considered an invaluable resource to be managed like various other resources and should immediately contribute to the achievement of organizational goals and objectives. An infrastructure element that allows certain important and relevant data to be transacted, prepared to provide content and info services that the user can use. It is important to create certain minimum socio-technical requirements for a resource qualifying element.

This book covers the unifying methods of the semantic web and the approaches for resource description to incorporate the various practices and traditions of information and knowledge organization. This book is an important book for information professionals who learn how to use the latest metadata tools and skills. It is also essential for LIS students taking courses in information organization.

Table of Contents

Chapter 1

Integrated Resource Discovery and Access of Manuscript Materials: the User Perspective

Jan Roegiers

INTRODUCTION

Recent developments in ICT have given hope to users of manuscript materials that some of their old problems will now be solved. Their primary question is possibly to be understood by the librarians and archivists who, more or less jealously, keep the treasures they are interested in. Strangely enough, the world of users is very often not so familiar to them. The worlds of librarians and archivists often differ more by the methods they use, than by the material they manage. To know which manuscript materials are kept by who is not always simple. National traditions, the fortuities of history and legal regulations have produced intricate situations that cry for better cooperation between those two worlds. According to the manuals used in the education of librarians or archivists, the definition of 'archives' is clear and unambiguous, but if you compare the manuals used in different countries, you observe fundamental differences, even contradictions. The best-known example is what is mostly called in English, *private papers*, in German *Nachlässe*, in Italian*spogli,* but in French, Dutch and other languages you read *archives privées* or something similar. In fact these collections of letters, personal notes and other documents received or written by a single person and kept by him/her, are not considered archives in the true sense by most nineteenth-century archivists and in the Anglo-Saxon tradition, where 'archives' is synonymous with 'public papers' or better 'public records'.

Consequently, you rarely have to look for private papers in British archival institutions, but in libraries. In most cases the content of the collection is described according to the rules of manuscript cataloguing, whereas in those countries where private papers form an important part of archival collections, they are described according to archival standards.

Description Methods: Bottom-Up Or Top-Down

The main difference between the description methods used by archivists on the one side and librarians on the other is the existence or non-existence of a hierarchical relationship to other items or other descriptions. Librarians got their inspiration for manuscript cataloguing from the description of printed material. Their main problem was the lack of standardisation of the manuscript material itself, for which they had to establish themselves a standardised terminology, for instance by attributing titles to works without any title page, to whole volumes without any label or inscription, or by attributing a uniform title to a very well-known text that appears under a different title. But they always describe manuscripts as individual single units, only linked to each other by the fact that they belong to the same owner and appear in the same catalogue, composed according to the same rules. This is a bottom-up approach, where the catalogue as a final result is the sum of non-related items.

Archivists, at least in our days, work top-down. They first describe the archive formed by a single institution, group, family or person, as a whole. This is the true unit of description. If possible, this description on the macro level is enriched by a further analysis of the archive, describing the series and finally the individual documents that compose the archive. In this way every element is described within its context, reflecting the original function of the document. The reader uses interrelated descriptive information, presented according to hierarchical relationships. Nowadays ISAD(G) offers a full set of standard descriptors that make it possible to computerise the archival description and to create the hierarchical relationship that is needed (ISAD(G), 2000). [1] Programs based upon EAD, a XML-DTD, make it fully possible to obtain from the so created databases automated inventories or lists that reflect the approach of a professional archivist, taking into consideration the hierarchical relationship of descriptions. Especially American professionals are aware of the fact that this system is not only useful for archivists, but also for manuscript librarians keeping ensembles of interrelated documents, in fact keeping archives or fragments of what originally were archives. The EAD approach has also found its way into the world of museums, where it proved to be the right answer to specific needs.

The top-down approach of the archivists is the result of an originally painful experience. In the seventies and eighties archivists could only envy their colleagues, the librarians, who successfully set up large automated catalogues of their complete holdings of printed material and who started linking the databases they had created. The first attempts to come to automated access for archival collections mostly failed or were, at least, disappointing. As the main reason archivists originally saw the lack of standardisation in their terminology and procedures, a consequence of the lack of uniformity in the material they wanted to make accessible. Once the problems of standardisation solved, the problem remained. Only in the early nineties they realized that the main problem was the inability of the existing programmes, derived from library software, to deal with the problem of interrelation of the items described. Only when they exchanged the bottom-up philosophy of library catalogues for a consequential application of their own top-down approach, reliable archival software became a reality.

The World Of The Reader

The third world is the world of the reader, in most cases not educated as an archivist or a librarian and nevertheless in quest of information in our treasuries. The questions of the users of manuscript material sometimes differ from those most librarians are familiar with. Often readers seem more interested in form than in content. They are interested not in the writer, but in the scribe; not in the text, but in writing errors; not in the book, but in the binding; not in how it is now, but in how it was centuries ago; not in how it is kept now, but in the previous owners. Traditional finding tools offer valuable information about these aspects and most readers are familiar with these instruments, the printed or manuscript catalogues, lists and inventories of the collections kept by libraries and archives. The number of the existing tools is considerable. Our Leuven University library keeps some 6,000 volumes of printed catalogues of manuscript collections and the collection of printed inventories of only Belgian archives numbers almost 2,000 items. Many of these instruments are rare to find and, if our collection of manuscript catalogues is rather important, our collection of inventories of foreign archives is very modest and almost limited to the neighbouring countries. For most countries there are no instruments that present a full bibliography of the existing instruments and to learn for instance whether the private papers of someone still exist, where they are kept, if there is full description of their contents, if they are accessible or under which conditions, can take years to find out.

Catalogues and inventories, if they exist, are very often the result of highly specialised knowledge and an enormous effort. Nevertheless they are mostly used by the readers in a rather primitive way and often there is no other way. It's incredible, but there are many catalogues of manuscript collections without any index, and if there is one, it is often limited to the authors whose work is included, although the catalogue itself mentions also e.g. the provenances, for many readers a more important fact. Similarly archival lists or inventories often present no index at all and if there is one, it is mostly limited to the names of persons and places mentioned. If there is one, readers mostly restrict themselves to a consultation of the index and rarely read the introduction where the arrangement of the list is explained or where they can learn about the origin, scope and limits of the collection. In many cases, however, there is no other way, even for the most experienced specialist, than reading the whole catalogue or inventory to find out if it describes something useful. Especially occasional users of archival tools are not familiar with the methods of description archivists use, although their way to present interrelated information asks for some specific effort - often remunerated by a rich catch.

The number of existing instruments is impressive and the quality of many of them is, according to their own standards, very high. But how to use them? Their mere number often presents such an inconvenience to the reader as making it impossible to start his research. Instead of running from one library to the other, glancing through some thousands of volumes to find out if they contain a copy of the text he is studying or a letter from the hero of his tale, he wants to dispose of instruments that will enable him to browse through endless amounts of descriptions, without leaving his desk. In the meanwhile, instead of going through the existing catalogues, he writes a standard letter to some hundreds of libraries and archives, asking if their collection might keep some documents he could use. Conclusion of this description of the reader's world: the reader wants more, not only in quantity and quality of description, but also in quality of retrieval and access. This only seems possible by new tools that take into account the many questions of the reader, and the many problems that derive from the material he is looking for. He doesn't look for a book that was printed in some hundreds or thousands of copies and of which he wants to know where the nearest copy is to be found, nor for a photocopy of an article that appeared in a journal, present in so many libraries. His first question is if there is anything that could respond to his needs and where he could find this rare bird, unique by definition. And finally, he has the same wish as all other readers: *tout, et tout de suite* ('everything, and immediately'). Until recently, this seemed impossible for manuscript material. Nowadays the reader, familiar with ICT, keenly awaits original solutions that will realise his old dreams.

I am convinced that the only way to proceed, to obtain rapid and lasting results, is to adopt the top-down approach of the archivists.

The Top-Down Approach

The bottom-up method of the librarians would mean that you could give an answer to all questions by introducing as many descriptions of individual items as possible into a common database, or by creating links uniting and covering several databases with individual descriptions. But what is an individual item? I can imagine how librarians, keepers of manuscripts, could describe all codices in their collections. But could they also describe all individual letters, kept among nineteenth and twentieth century private papers? And are they aware of the millions of letters and other individual documents, kept in an average archival depository? Or are they going to select between documents worth description and others that are not? This is why I don't believe in the final success of famous enterprises as the Dutch CEN project (*Catalogus Epistularum Neerlandicarum*), aiming at a full description of all individual letters by Dutch-speaking authors since 1600, regardless of their present depository. [2] It is a typical project drafted by librarians who aren't aware of the treasures kept by archives and who forget that the normal place to find letters is not a library, where they are mostly kept as individual 'autographs' and sometimes without specification of the addressee, but an archival depository, where letters appear in their full context and where you are aware that the addressee is often as important as the writer of the letter, sometimes even more. A bottom-up approach is only valid for very important and highly valuable materials, such as illuminated manuscripts and other medieval codices, for letters by Erasmus and Thomas More, or for the private papers of individuals from the category of Goethe, Einstein, Proust, James Joyce, Picasso or Manzoni, people of whom you know that every scrap of paper will be published and commented on for centuries. It seems very typical that the bottom-up strategy is used by the MASTERproject, short for 'Manuscript Access through Standards for Electronic Records', a project funded by the European Commission with De Montfort University of Leicester as its most active partner. Their goal is a single online catalogue of medieval manuscripts in European libraries.

The top-down approach means that the first task of librarians and archivists would consist in establishing an easily accessible database, presenting an overview of all existing repositories where manuscript material is to be found. What the user really needs is *something abouteverything,* not *everything about something*. I cannot enough insist on this principle, so often neglected by perfectionist librarians and archivists. The first need of the reader is to be aware of the mere existence of resources that could be useful to him. A very short and simple description is often enough. To put it in good scholastic Latin: *melius est esse quam non esse*, or: it is better to be than not to be. And in our days, as we see every day in our contacts with students, not to be on the Internet equals not to be at all.

It makes sense to plan the general overview as a national task for every European country and to link these databases at the European level in a later stage. Austria has set a good example with its *Handschriftbestände in Österreich,* set up by the Kommission für Schrift- und Buchwesen des Mittelalters of the Academy of Sciences (KSBM), the UK with the *National Register of Archives* (NRA),maintained by the Historical Manuscripts Commission (HMC). In fact, similar projects are set up in several countries. In my own country, I can mention the Archiefbank Vlaanderen, meant to become a national register of private archives and shaped after the British model. In many countries it could make sense to think of two databases, one for libraries and another for archives. Nevertheless the distinction becomes more and more artificial, especially since the creation in many countries of specialised documentation centres that collect printed and manuscript material, archives as well as books, periodicals, posters or photographs, all related to a specific topic. You also have to take into account that other types of institutions very often keep important collections of manuscript material, such as museums. Of course it becomes also possible to create a national or regional instrument that enables cross-searching in different databases, such as is done with the DutchCultuurwijzer, where the holdings of archives, libraries, museums and other institutions can be searched simultaneously. A very successful example of such a tool is the Online Archive of California (OAC) that provides access to materials such as manuscripts, photographs and works of art held in libraries, museums, archives and other institutions across California.

Apart of general information on the library or other institution, the database should comprehend as soon as possible an overview of all existing collections, holdings,*fonds, Bestände*, or how they might be called, presenting a description on macro-level and references to the existing catalogues, lists, or inventories, published material as well as unpublished. How can one expect a scholar who has never visited the Brussels Royal Library and who discovers that some ten thousand manuscripts of this library are described in a published catalogue in thirteen volumes, [3] without any kind of index, to know that he could find in the Library itself a full index in the form of a card system? It might be worth travelling to Brussels, or writing a simple request for information, instead of going through thirteen volumes. The presentation of individual institutions should imply a standardised typology of the depositories themselves, making it possible to link similar institutions as is already done by the Deutsche Archivschule in Marburg which can show you the way to University Archives all over Germany.

Integrated Discovery & Access

In a further stage the many already existing tools should be fully integrated into the databases. The splendid catalogues, lists and inventories could be integrated in many ways. It would be a great loss of information if they should be replaced by uniformly standardised descriptions, made according to some MARC standard and considerably poorer than the products of the great scholars who composed the catalogues we are familiar with. The most attractive of all solutions, for the experienced user as well as for the beginner, is a scan of the full text, enriched by tagging according to something as XML, for instance by using the EAD-DTD. This would enable systematic searches through different catalogues, and at the same time keep all less standardised information. Something similar is set up by the British Library for the automation of the 70 volumes of its manuscripts catalogue, describing a million of manuscripts. France has been inspired by this example for the automation of the more than hundred volumes of the *Catalogue général des manuscrits des bibliothèques de France* (Creff, 2001). I hope that this work in progress will result in a database as rich as the printed edition. It is likely that the major libraries in Europe dispose of the means for a similar conversion project. The great question is who will take care of the conversion of the other, smaller libraries, very often great in their collection, but poor in their present financial and personal means.

Unfortunately much manuscript material, kept by important as well as by smaller libraries, archives, documentation centres and other institutions, such as museums, has up to now not been described at all. As I have already mentioned, it is much more important for the reader to know that a collection exists and to have a general idea of its content, than to dispose of a description in full detail of every single item. In a second stage short descriptions, according to a generally accepted standard, could enrich the macro-description at low cost. It also seems preferable to use the same standard to encode the essential data contained in old manuscript lists or in rather administrative inventories and acquisition lists. In a top-down philosophy it remains possible at any moment to enrich the existent description with more details, e.g. by analysing the letters or other individual documents united in an item originally described as a whole, or by describing the single illustrations and by adding information on other aspects than the content itself.

A full scan of the document itself can of course be very useful for the reader, but is not what he generally expects. It makes sense of course, out of reasons of conservation, for extremely precious or fragile materials, for documents that are frequently used, for documents that are of more interest for readers abroad than for a native public or for documents with a high pedagogical value. It is unthinkable to scan completely the vast collections of post-medieval times. Readers often regret the enormous sums absorbed by prestige projects for reproduction or scanning that merely serve the image of the institution and its keeper, whereas large parts of the holdings have never been described or remain inaccessible for various reasons.

Manuscript material often asks for descriptors that are very unfamiliar to librarians, used to printed material. Some examples. First there is the problem of accessibility, often depending on a special permission, especially for more recent documents. It is very important to the reader to be aware of these restrictions. The more he gets easy access to faraway collections, the more it becomes necessary to avoid disappointments, by warning him before he takes the plane. The same is more and more true for copyright problems, where legal regulations become more and more strict and lawyers more and more inventive to enrich themselves. There are also other unexpected elements that can interest a reader of unique material. When I work at the Handschriftensammlung of the Österreichische Nationalbibliothek, I find in every volume a slip bearing the names of the earlier readers who used it. Of course, it is always flattering to know yourself the successor of a famous scholar, but often it is also very useful to know that you are or possibly are not aware of the earlier and especially the more recent use of the document. We have to avoid cases such as the one of the unfortunate scholar who recently presented a manuscript for publication to me, a fully annotated edition of a text from the Vatican Library of which the deciphering and transcription had taken almost a month, and whom I had to tell that an edition of it had been published a year before. Libraries and archives that keep unique material have to establish good contacts with their readers, to be informed about work in progress based upon their collections, and to use this information when necessary. The use of a unique document creates a link between the reader and the document, which has to be recorded, and, as far as our susceptibility for privacy permits, could be communicated to other readers.

A question I have avoided until now is in how far an integrated form of access for printed and manuscript material meets the expectations of the user. Let us say first that integrated access of manuscript material, kept by several institutions, is one of the reader's main wishes. The present depository of a document is often so fortuitous and unexpected, that integration of the existing or new finding tools is an absolute demand. The first requirement for such integration is the use of good authority files for persons and corporate bodies. Once more archivists have been the first to realise that this is not only a question of avoiding homonyms by the use of standard names, but also includes a set of descriptive elements that have to be presented according to a specific standard. Because of the importance of access points of archival retrieval, the Committee on Descriptive Standards (CDS) of the International Council on Archives (ICA) developed a separate standard, ISAAR(CPF), short for 'International Standard Archival Authority Record for Corporate Bodies, Persons and Families'.[4] This example has inspired the initiators of the LEAFproject, 'Linking and Exploring Authority Files', supported by the European Commission and aiming at a 'pan- European Central Name Authority File', meant among others to serve for MALVINE. It is evident that it doesn't make sense to use different authority files for printed and manuscript material. If LEAF really becomes a common European tool, it will enable integrated access of materials of different types and the readers will surely appreciate the improved extended possibility for retrieval. Libraries that would link their catalogue of printed material to LEAF and enrich LEAF by their own authority files would be of a great help for their own and other readers.

CONCLUSION

MALVINE is meant as the answer of the library world to the need of the users for integrated discovery and access of modern manuscript material (Weber, 2002). It is described as "an electronic network of European institutions, independent of heterogeneous technical solutions, to enhance access to disparate holdings of modern manuscripts and letters, kept and catalogued in European libraries, archives, documentation centres and museums". This multilingual metadata based search engine for a specialised sector has to provide harmonised access to a large number of European collections. The list of participants in this stage of the project is already impressive. MALVINE is inspired by the archival EAD standard. The ambitions and the expectations are high and the first results, visible since July 2003, are promising. In my opinion the real success of the enterprise will depend on the ability to adopt a top-down strategy. Otherwise the content of MALVINE will remain an arbitrary accumulation of knowledge already available on paper. Another question that is not fully clear to me is the assignment of tasks between MASTER, meant for medieval manuscripts, and MALVINE, for post-medieval or modern material. There is a large category possible in between! The success of MALVINE and similar projects will also depend on the ability of librarians, archivists and others who keep the keys of the treasuries, in explaining the top-down approach to their readers. Readers are used to the bottom-up approach of library catalogues and only a few of them who are experienced in intensive use of archives, are familiar with the other way. Interfaces will have to focus on this pedagogical aspect. If they succeed, the readers will finally have the same access to manuscripts as they now enjoy for printed materials. Readers will no more have the impression of belonging to an underdeveloped third world. The three worlds of librarians, archivists and readers will meet.

REFERENCES

1. Creff, Jean-Arthur. « Quelle informatisation pour le catalogue général des manuscrits des bibliothèques publiques de France? ». *Gazette du livre médiéval*, 39 (automne 2001), 41-45.

2. Dongelmans, B.P.M., A.M.T. Leerintveld. *Digital access to Book Trade Archives*, Leiden : Academic Press Leiden, 2002. VIII, 84 p.

3. Weber, Jutta, "MALVINE, LEAF and Kalliope: Some co-operation models". In:*Digital access to Book Trade Archives*, Leiden : Academic Press Leiden, 2002, p. 49-68.

NOTES

[1] ISAD(G) is a result of the Committee on Descriptive Standards (CDS) of the International Council on Archives (ICA). A second edition was published in 1999 and is available in English at http://www.ica.org/biblio/cds/isad_g_2e.pdf and French at http://www.ica.org/biblio/cds/isad_g_2f.pdf

[2] The CEN catalogue can be consulted via the catalogues of the Royal Library at the Hague, on the spot or by registered readers. The project can also be regarded as merely a catalogue of the holdings of a (large) group of libraries; in this sense it is rather a success, but can be misunderstood by readers who see it as a database of all relevant materials.

[3] Joseph van den Gheyn a.o., *Catalogue des manuscrits de la Bibliothèque royale de Belgique*, 13 vols., Brussels 1901-1948.

[4] More information and the full text of the latest version can be found on the website of ICA. A group of archivists, who met for the first time in Toronto in March 2001, created a more complete high-level model for the recording and exchange of information about the creators of archival materials, termed "Encoded Archival Context" (EAC), compatible with the ISAAR(CPF) standard.

Chapter 2

Staying Relevant in the Google Age: Implementing Vertical Search at the University of Manchester — a Technological and Cultural perspective

Lorraine Beard, Andy Land

The John Rylands University Library, The University of Manchester, United Kingdom

The John Rylands University Library, The University of Manchester, United Kingdom

ABSTRACT

John Rylands University Library at the University of Manchester makes a clear commitment to improving resource discovery in its New Directions strategy. In addition an external review of the Library's systems portfolio identified a clear need to improve the user experience in finding and accessing information. There are also drivers to ensure better utilisation of the Library's investment in e-resources, as well as the need for the Library to respond to the question of how libraries continue to stay relevant in the Google age. Taking all these factors into account the Library took a strategic decision in 2010 to implement a vertical search solution and it established a formal project to do so in Summer 2010. The project has involved working in close partnership with a number of key stakeholders from the Library, IT Services, multiple vendors and the academic community and is a good example of the changing way in which libraries need to plan and manage their projects in order to improve and deliver services which address the needs of their users. Effective governance, communication, change and project management have been key to the success of this project and highlights the growing importance of these skills in libraries today.

The paper will consider the complex technological issues around integrating data from different sources and using different metadata standards. The project has involved vendors working co-operatively together to enable data to move between systems and different metadata formats, which also presents challenges. The cultural issues involved in implementing vertical search will also be discussed, including a shift in the Library's internal philosophy towards search, the possible impact on the traditional Library catalogue and associated workflows and the changing user perception of the Library in response to vertical search.

The project is an excellent case study in new ways of working for libraries, tackling complex technological issues and remaining focussed on improving user experience. The paper will be of interest to other libraries considering or in the process of improving their library's resource discovery experience

Key Words: Vertical search; research libraries; resource discovery

INTRODUCTION

In the John Rylands University Library New Directions strategy there is a clear commitment to improving the search experience. In 2009 the Library conducted an external review of its systems portfolio, to ensure its systems where aligned to its strategic ambitions. One of the main recommendations of the review was that due to the diverse range of interfaces provided by the Library for its users, that we should simplify the user experience by implementing a vertical search solution. This would also provide greater value for money by increasing discovery of the resources that the institution has made a significant financial investment in. A project was therefore commissioned to implement a vertical search solution.

Project Implementation

A project board was established with senior Library leadership and representation from across the Library as well as academic representation. The entire project covered the period from February 2010 to February 2011 and involved procuring and implementing a search solution, building a portal interface to the solution and embedding it within the Library's web presence, as well as decommissioning any systems no longer needed. To start the procurement process the requirements were defined, a market survey was conducted and we invited a number of vendors to give on-site presentations focussed on our requirements. From this process a product was selected in August 2010, Primo from Ex Libris, and a contract signed by October 2011. An implementation team was established with membership consisting of the different system owners and a Project Manager assigned from ExLibris working with a local Project Manager. The team had an initial kick-off meeting and then had a weekly conference call throughout the project. The team focussed on the data migration from each of the datasources being integrated into Primo, agreeing the interface and ensuring integration with existing University systems. The data sources that we focussed on importing included catalogue data, e-resources data, the institutional repository Manchester eScholar, LUNA (image database) and Elgar (archive database).

We opted for a cloud-based solution so any infrastructure issues were managed by ExLibris. This solution fits in with the wider University IT strategy of using cloud and virtualisation to reduce carbon footprint and also means the total cost of ownership was reduced by requiring less staff support. The main concerns with this were related to reliability of a hosted solution, but so far this has not been an issue. Also access to the back-end is limited as we are sharing server space with other institutions. However, the Primo APIs are good, so, thus far, this has not been a problem. One issue is that we have not been able to get a Manchester specific URL (http://man-fe.hosted.exlibrisgroup.com) — we have a bridging page (http://www.library.manchester.ac.uk/searchresources/librarysearch/).

We met our implementation timescale of going live by February 2011, with our implementation marketed as Library Search. Our experience of working with ExLibris and the project manager assigned was extremely positive.

Systems Integration Issues

Talis

ExLibris had not integrated Primo with the Talis products before this project and so to integrate with Talis Alto Ex Libris built a new pipe based on MARC21. We have not yet achieved real-time availability information in Primo, although this is expected shortly with Talis and Ex Libris working together to achieve this. Talis Alto tables are complex and the existing Talis export routines did not grab the appropriate data and supply it in the required way. Ex Libris worked with Talis to resolve this and a new routine is now in place. For updates Manchester have created an sftp area where we place a nightly update file which is then grabbed by Ex Libris.

We are also working to achieve integration with Primo and Talis Aspire, our reading list solution, so that content bookmarked in Talis Aspire from Talis Prism imports key bibliographic data into Aspire.

Manchester eScholar

As eScholar is built using Fedora Commons it is able to supply data in multiple formats. It was agreed that data would be supplied in MARC21 as a pipe already existed. However, a lot of the metadata in eScholar does not easily map to MARC21 and therefore we agreed that only the key bibliographic data be exported to Primo. Once a record is found the user can then view the full eScholar record as a reformatted eScholar page within the search results record iFrame.

Primo works on the premise that a record will either relate to a physical item held by the Library or to an electronic item that it has access to. In the case of eScholar this is not always the case. The majority of records on eScholar record the fact that a member of Manchester staff has produced a particular piece of work but does not provide access to this. We are still working with Ex Libris to provide a helpful way of indicating this.

LUNA

LUNA is not yet currently fully OAI-PMH-compliant and therefore Ex Libris are not currently able to harvest our LUNA data into Primo. This is recognised by LUNA Inc. as a problem with their OAI-PMH implementation and they are working on a fix which will be applied in the next release (July 2011). There is, therefore, no LUNA content on Primo at present. LUNA metadata is based on extended DC and a significant amount of work had to be done to map the LUNA metadata to the fields used in Primo (both for filtering and search purposes).

User Interface Issues

Primo comes with an out-of-the-box interface which, if you use a hosted service, can be revised in a limited number of ways (addition of logo, amended field labels, changes to the positioning of navigation). Ex Libris were able to quickly provide an initial interface with some content only three weeks into the project. The search window has been embedded into the Library's website homepage, and into the University's staff and student portals.

We have a mobile-friendly version of Primo developed which is waiting to be launched alongside the University's mobile service.

Most interface issues have been minor and easily resolved. There remains a problem displaying non-roman characters in some Primo Central records caused by metadata supplied by some publishers — this is being addressed on a case by case basis by Ex Libris. There also remain issues with the date-newest sort Which, again, has been diagnosed as a problem with some Primo Central content and is being dealt with by Ex Libris.

The grouping of editions using FRBRisation works well on the whole. However, the display of the latest edition of the book, whilst making sense for standard course materials is not always appropriate for Special Collections materials where latest is not always best. Another issue is that filter options in Primo are driven by the metadata in the retrieved records. For FRBRised records only metadata from the record been displayed will be included in the filter options. One could, therefore, have an older edition of a book available at a different location but this location would not appear in the location filter option.

The Primo results page uses iFrames to display content from other sources such as the Library catalogue. We have, though, found that LUNA images do not display well (the ability to zoom in and out does not work well in an iFrame), so we have LUNA records opening in a new window.

Authentication Issues

Integration with Shibboleth authentication was straightforward to achieve. The outstanding issues are that if a user comes to Primo via the University portal having already authenticated, they are not immediately recognised by Primo. Also, if users authenticate in Primo and then go to the Library catalogue, their credentials are not passed through. ExLibris and Talis are working on this together.

Search

Primo uses the Solr search platform.The decision was taken during implementation to provide very few pre-search filter options (you can only choose to exclude journal articles) and to focus instead on post-search filtering. This has been an issue for some users (mainly Library staff and postgraduate researchers who want to pre-select the databases they are searching). The speed of searching and the returning of results is well within what could be considered reasonable. Feedback from Special Collections staff suggests that we need to retain the option of browse searches by title and author and also to be able to turn off the grouping by edition and version (see above).

Management Information

Primo does not make use of Google Analytics but there are management information tools within Prism. Usage of Primo has been high: in the first month of launching there were 122,427 searches and this increased in March 2011 to 205,445 searches.

Cultural Issues

After the implementation, we have surveyed students and staff to gain an insight into their initial user perceptions and their perceptions of the solution after one semester of usage.

To do this, we conducted an email survey of our student ambassadors asking the following:

- What do you like about vertical search?
- What do you dislike?
- How can we improve?
- How does it compare with Google and the Library catalogue?

We received a 9.5% response rate to the survey, with the following results:

Have you used Library Search? The results are shown in Figure 1.

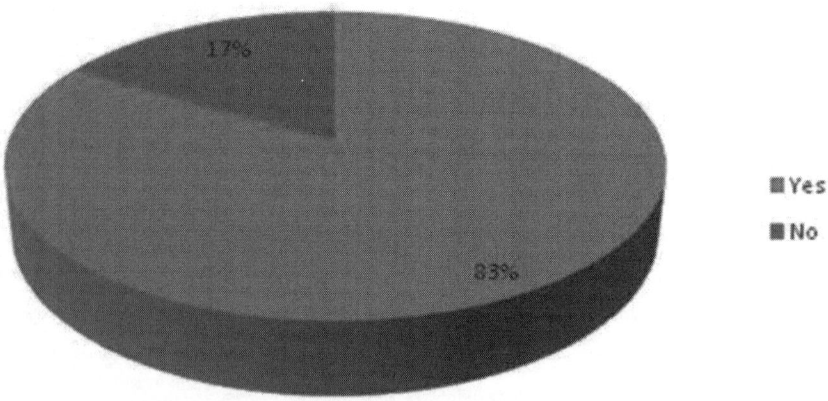

Figure. 1 Student usage of Library search at the University of Manchester, from February to June 2011: 83% said they had used Library search while 17% said they had not used it.

How simple was it to use? The results are shown in Figure 2.

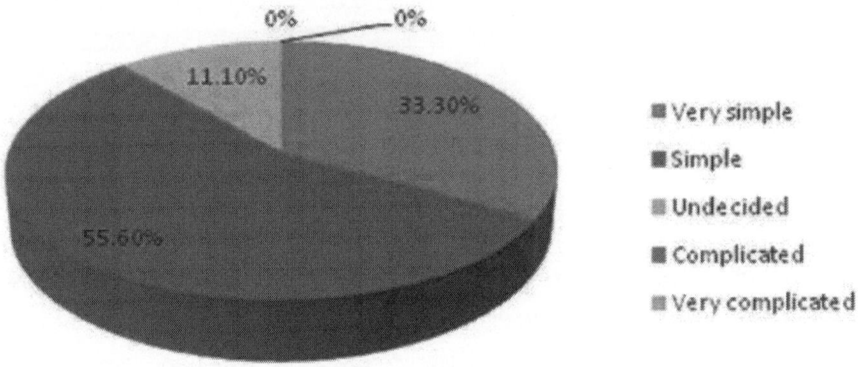

Figure. 2 Feedback from students showed 33% said it was very simple to use, 56% said it was simple, 11% were undecided. Three people skipped the question entirely.

Did you find what you were looking for? The results are shown in Figure 3.

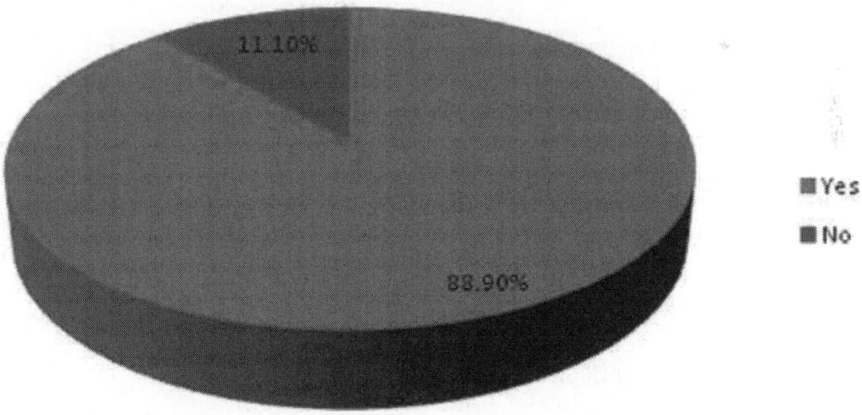

Figure. 3. Student feedback on whether they had found what they were loooking for using Library search: 89% of respondents found what they were looking for; 11% did not.

A summary of the free-text student feedback is given in Table 1.

Table 1. Summary of student feedback on Library Search.

Likes	Dislikes
Fast	Too many results
Easy to use	Unrelated materials
Journal access	Difficult to locate a particular paper
Very clear	Not all books have images

In terms of improvements to Library Search, the main improvements requested are: more images and extra search options. In comparison to Google and the catalogue students felt that Library Search is faster and simpler and that Library search is better for in-depth search, with Google better for quick background search. Some students commented that they could 'Find journals I can't find in Google in Library Search'.

Library Staff Feedback

To gather feedback from Library staff we asked the following questions via email:

- What were your expectations?
- What were your concerns to moving to vertical search?
- What has been the response from schools?
- Benefits and drawbacks
- What worked well/less well with the project?
- What needs to improve?

Expectations prior to implementation were mixed, from pessimistic to excited to have a good way to compete with Google and that it would produce more relevant results and look and work like Amazon. The key feedback from Library staff regarding Library Searchisshown in Table 2:

Table 2 Summary of Library staff feedback on vertical search.

Positive Feedback	Negative feedback
• Faster	• Students need to understand it has limitations
• Easier for students	• Too much content
• Promotes visibility of repository content	• Facets have variable value
• Range of resources	• Different results to catalogue
• Alerts valuable	• Range of resources
• Preferable to Google	• Not suitable for researchers
• Less intimidating	• Not suitable for Special Collections material

In terms of improvements that staff hoped to see, these were in the areas of improving content coverage, particularly in the field of business and law, improving search including Boolean searching, improved search filtering and search history. The next area for improvement was integration with more resources including LUNA, Elgar and Aspire. There were also comments about better communication and the need to manage user expectations by giving guidance on how Library Search can be used.

Lessons Learnt — Technical/Implementation

The main lessons learnt were around the implementation schedule. We were a little ambitious with the timeframe for the implementation and this limited the amount of time that was available for both usability testing and the metadata mapping which is crucial to the effectiveness of the search results. In order to work to a tight implementation timescale we chose to exclude some resources from the original project and the contract with the supplier and instead would integrate them later outside the project window. This presents difficulties for users in selling a solution as providing full coverage. In retrospect it would have been better if we could have extended the project timeline in order to implement as many data sources as possible to be able to launch a fuller solution. Launching at the start of the academic year would have perhaps also had a more favourable response for users.

In terms of the cultural implications with Library Search there have been a number of key lessons that have emerged. It is clear that Library search is unlikely to be the one stop search shop that we were hoping for. In some subject areas key resources are not available within vertical search and for researchers the subject-specific search functionality that cannot be achieved through vertical search is essential. We therefore need to provide clear routes to other sources of information and guidance on which works best according to the user's requirements.

Above all we need to engage closely with users as vertical search continues to develop, in order to map search habits and desired outcomes to the interface and functionality we present. We cannot make any assumptions about how users prefer to search without extensive and iterative user testing. One very clear outcome was the need to invest time with Library staff throughout the implementation process and beyond to manage their expectations of vertical search and to gain their valuable professional insights. We had a small steering group which did provide this input, but ideally we would like to have invested more time in the project to get the buy-in from a wider range of staff before the product was launched. A strong marketing and communications plan would also have helped to deliver a clear message about what vertical search is, how it fits in with the search landscape generally and what its limitations are. We opted instead to have a soft launch of vertical search which has left a number of unanswered questions for some users.

The Future of Library Search at the University of Manchester

Firstly we would like to integrate more databases into Library search and complete the integration of LUNA and Elgar. We would like to improve the search functionality by including more prefiltering options. We would like to run some detailed user focus groups to establish what our priorities are for future development and to review the usage patterns of Library search compared to Google, the catalogue and subject databases, in order to inform how we market Library Search most effectively. Key groups of Library staff that we need to obtain more buy-in to the vertical search are the subject librarian team and librarians in Special Collections. New developments that we would like to consider for the future include a recommendations tab and the use of user created content.

SUMMARY

In response to the question originally posed of whether vertical search stays relevant in the Google age, the answer is it is a positive development for us in improving the user experience, but there is still a long way to go before vertical search will be embedded in search habits as deeply as Google is. It does however demonstrate the Library's willingness to innovate and improve services in line with customer expectations. For Manchester we now intend to engage closely with a range of customers, including Library staff, to improve the current implementation to deliver better results and a better customer experience.

REFERENCE

1. John Rylands University Library New Directions strategy, http://www.library. manchester.ac.uk/aboutus/strategy/_files/new-directions-2009-2012.pdf (2009).

Chapter 3

Hakka Genealogical Metadata Principles and Practicalities

Chen-Chi Chang*, Ruey-Ming Chao, Ting-Yu Wu

Institute of Information and Society, National United University, Miaoli, Taiwan

ABSTRACT

Genealogy (Zupu in Chinese) is the history of families. Genealogy has been widely used for the tracing of their lineages and history. Hakka is a special Chinese ethnic group. To explore the origin and history of Hakka people and their culture, this study starts with the genealogy. After several Hakka genealogies collected, the content analysis is performed in this study. According to the qualitative data analyses, this study designs the pedigree metadata standards for Hakka to implement the union catalog. The Hakka Pedigree Metadata is comprised of many individual family and ancestors, each of which has a process in place for archiving and distribution of genealogy. Hakka genealogical metadata contains the following types of constructs. 1) Pedigree: use given name and lines expressed the genetically related. 2) Individual: describe the personal information. 3) Family History: the story of the ancestors is narrated. 4) Bibliography: description of the genealogical book. 5) Miscellany: included the unique documents that could not categorized in the above four classification. Results of this study can be applied to genealogies integration and implement the union catalog.

Keywords: Hakka Genealogy; Genealogy; Metadata; Pedigree; Pedigree Chart

1. INTRODUCTION

The language from the ancestors is the symbol and spirit of Hakka. Hakka language as the fundamental of Hakka culture, genealogical documents is important archives that recorded the ancestors' words, family instructions and surname is particularly important. Hakka is a special Chinese ethnic group. The Hakka people are one branch of the Han Chinese who lives predominantly in the provinces of Taiwan, Guangdong, Jiangxi, and Fujian in China. There is a proverbial saying in Hakka that, "It's better to sell one's ancestor's field than to forget one's ancestor's language; it's better to sell one's ancestor's house than to forget one's ancestor's voice". The concept of ancestor is really important for Hakka family. It makes the main barrier as the scholars doing the related research. The style of writing have different genealogy wording owing to the space and time difference, so there is no standard format. Several studies indicate that genealogy plays an important role in academic and social science [1,2]. Genealogical records related to the precious historical data of family. There are importance of historical value include economic, demographic, educational, ethnic and religious, etc. Genealogy (Zupu in Chinese) is the history of families. The genealogy is the historical trajectory of the family. The genealogy represents one of ethical rules for human survival and development. Recent years have seen increased attention being given to family tree writing in the Hakka genealogy literature. To explore the origin and history of Hakka people and their culture, this study will start from genealogy. Hakka genealogy doesn't have a writing format standard, so the preservation of Hakka Genealogy loss critical information. Scholars encounter certain degree of difficulty, who likes to collate these genealogies. A complete view of Hakka cultural characteristics is necessary. Therefore, establishment of the Hakka Pedigree Metadata is considerable. Well-designed Hakka Pedigree Metadata makes it is easier to merge genealogy and provide offspring easy to find their ancestors. Relatively little discussion in considerable amount of literature on Hakka Genealogy, the status of collected Hakka genealogies is a fundamental issue. Recent years have seen increased collected being given to Hakka genealogies in the literature. As the lack of union writing standard, family trees cannot integrity smoothly. Because the family tree recorded in each genealogies are separate. First, this study explores the genealogy metadata standards of countries in the World. To understand the differences between various genealogy, and the pedigree metadata standards for Hakka to implement the union catalog as the reference standard. Second, characteristics of the Hakka Genealogy will introduce in this study. Design the pedigree metadata standards for Hakka is claim of Hakka culture. Hakka genealogies of writing can be more standardized.

2. LITERATURE REVIEW

2.1. Genealogy

Family history can be a very interesting and as a research field has not yet been much explored. The Genealogy is one kind of family history record. In previous study, Liaw (2003) provides extensive discussions of genealogy, overview of genealogical documents reorganization, antiquarianism, as well as the genealogy of the contribution. Liaw (2003) also noted that there are important historical value of genealogy include economic, demographic, educational, ethnic and religious, etc [3]. Nowadays, several studies indicated that genealogy as the critical reference because of their potential to disentangle complex population histories [4-6]. However, the genealogy is a gray literature (non-conventional literature). It can't be obtained through purchasing channels. To collect these genealogies, it need people to provide or donated voluntarily. Field investigation is another way to collect the genealogical data. In the past, genealogy often consists of a family or a clan as a unit and writing in the paper. Various social and environmental factors, such as war, natural disasters and migration lead genealogies has been damaged or disappeared. In order to save and protect the genealogies, it is necessary to make these genealogies in the digital form. In the Western, the family tree is expanding pedigree in graphic form. The family tree chart presents a different history and cultural connotations that consisted of religious and secular. Family tree is one of documentation and digitization of genealogical to present. The genealogy employed the family tree as the structure for family members. The graphic family tree chart linked according to the genetically relationship of family member by information commutation technology. The family tree concept is similar to Chinese genealogy. Several sites, such as Ancestry.com [7], and Taiwan Genealogy Online [8] offer the services that genealogy writer and provide offspring to find their ancestors.

2.2. Genealogy Metadata

The regular definition of metadata is data about data. This explanation derived from the Metadata Workshop Report in March 1995. Renato Innella (1998) proposed a new concept of metadata is structural data about data. Metadata plays an important role not only in the digital library and museum environment, but also other areas [9]. For example, metadata is control of medicines management in the medical profession [10]. Metadata has different name in various domain fields. In library science, metadata is called Library Catalogue. In museum science, the metadata is called documentation and data standard [11]. Lorcan Dempsey and Rachel Heery introduced three types of metadata standards: simple formats, structured formats, and rich formats. Simple formats are that the data structure of a company exclusivity, emphasis on full-text indexing. Structured format is that standardized in formats. Characteristics based on the field as the main structure, such as, Dublin Core, IAFA the templates, RFC1807, SOIF, LDIF, etc. The formats design has a problem that lack of relational functions. Rich formats have been compliance with international standards. Its characteristics are detailed and complex flag to present the contents of the resource. For example,

ICPSR, CIMI, EAD, TEI, MARC, etc. [12]. However, Dublin Core is the most suitable Document-Like Objects. Dublin Core originated during the 1995 invitational Online Computer Library Center (OCLC)/National Center for Supercomputing Applications (NCSA) Metadata seminar. Dublin Core design is based on a simple description applies to different metadata standards [13]. Therefore, Dublin Core can also apply to genealogy metadata. There are many genealogy metadata formats at present. **Table 1** shows different genealogy metadata standards.

2.3. Hakka Genealogy

Hakka is a special Chinese ethnic group. Instead of living in the same region, Hakka people were connected by their culture, language and heritage. The Hakka people are one branch of the Han Chinese who lives predominantly in the provinces of Taiwan, Guangdong, Jiangxi, and Fujian in China [18]. There will experience many difficulties in migration of Hakka people. Hakka people have been consciousness that not afraid of predicaments and courageous spirit. The reason for Hakka people special attention to the genealogy records is that they leave hometown frequently. Hakka genealogy contents include migration process of the family, genetically related, marital status, Family Instructions, and so on. The contents of Hakka genealogy are inseparable from Hakka culture. If you understand Hakka culture, you will comprehend Hakka genealogy what elements is necessary Hakka culture has the following characteristics [19]. Hakka culture has the following characteristics [20,21]. First, respect for ancestors, mourn for ancestors, the lessons from the ancestors is the symbol and spirit of Hakka. Consequently, it recorded Family Instructions in the genealogy. Hakka people pay careful attention to own parent's funerary rites. Second, believe that mysterious power is the common tradition in the Central Plains. Hakka people believe that Feng Shui has a mysterious power and will affect their fortune. Third, this is a glorious to become officials in the early China. They think became officials can upgrade themselves status in society. If they don't have any ancestors of celebrities or aristocrats, it

Table 1. Genealogy metadata standards.

Metadata Standards	Introduction
GEDCOM	GEDCOM Stands for GEnealogical Data COMmunication. It is a Common file format that allows users to exchanges genealogical files between different Programs [14].
GedML	These pages describe GedML, a way of encoding genealogical data sets in XML. It combines the well-established GEDCOM data model with the XML standard for encoding complex information. The result is a representation that can easily be converted to and from GEDCOM, but can be manipulated much more easily using standard tools: notably, by using an XSLT processing such as Saxon [15].
GEDC	GEDC is an XML format for genealogical data. Although it's relatively new, GEDC has already stood the test of time. It has been in everyday use since early 2003, in parallel with its development, and its evolution has been guided by practical needs as they arose. GEDC is currently being used to maintain several genealogies as large as 25,000 persons. GEDC originated as an outgrowth of the GEDCOM 6.0 Beta spec from the Church of Jesus Christ of Latter Day Saints. It still bears a superficial resemblance to its predecessor, but many minor changes have been made as well as several major ones. Especially, GEDC has followed the lead of Gentech's Genealogical Data Model (GDM) and broken away from the mindset implicit in GEDCOM: that genealogical data is entirely consistent and correct [16].
GEDCOM X	GEDCOM X is partitioned into different data models that each supports a distinct aspect of the genealogical proof standard. Specific serialization formats (e.g. XML, JSON) are defined to describe the ways the data is written to a file or exchanged over the Internet. You may also be interested in learning more about how GEDCOM X is being extended to provide for specific use cases and requirements such as field-based image extraction and standard Web Service APIs [17].

humiliates them. It is shame at they don't have any ancestors of celebrities or aristocrats. Forth, patriarchal clan culture is unique culture of the Hakka. The extended family consists of a lots nuclear family. It will write on paper that the name of the ancestral hall. These points will affect the contents of Hakka genealogy. Genealogy writer will be carefully recorded that ancestors of celebrities, family instructions, Feng Shui and The name of Ancestral Hall. The nature of Hakka people owe to ancestors. It will affect the emphasis on their genealogy.

3. RESEARCH METHODOLOGY

To establish the Hakka Pedigree Metadata and merge genealogy, this study starts with the content analysis of genealogy. Both meta-analysis and content analysis were performed to gain the genealogy metadata standards of Hakka. This study also identified which surname is Hakka surname, and chooses for the representative of Hakka genealogy. Data were collected primarily from the Taiwan Memory database in National Central Library. National Central Library has rich collection of historic documents in Taiwan. Recent years, the rich range of historic documents on Taiwan all collected through a variety of channels. National Central Library entered into a cooperative agreement with Genealogical Society of Utah (GSU) in 2004. Also acquired copies of materials in microfilm collected from civil society in Taiwan in 1970 — there are contain a total of more than 794 microfilms and more than 9000 records of family genealogy [8]. In order to properly preserve these genealogies of Taiwan, National Central Library used digital technology. It makes the expedience as the scholars doing the related research. Scholars can study the genealogy by browsing features. When data is collected, Hakka surname as judgment standard by the expert opinions. Study on genealogies and the genealogies catalogs are classified. After the analysis and filter the unnecessary catalogs and Hakka culture catalogs has be reserved. This study designs the pedigree metadata standards for Hakka to implement the union catalog.

4. RESEARCH RESULTS

4.1. Hakka Genealogical Metadata

This research study collected related literature about Hakka surnames, concluded of the 28 Hakka surnames. The reference genealogy surname standards in accordance with 28 Hakka surnames. It had difficulty to obtain because genealogy is gray literature. This study only refers to 25 genealogies. In this study, reference 25 genealogies compiled pedigree metadata. In addition, the catalog of Hakka culture will be preserved. In this study, the genealogy of information is program to divide into five constructs, these constructs contains many sub-constructs. The genealogy of information is consists of pedigree, individual, family history, bibliography, miscellany. Key concepts are presented in **Figure 1**.

Genealogy of category contains the following types of constructs. Pedigree: Use given name and lines expressed the genetically related. Individual: Describe the personal information. Family History: The story of the ancestors is narrated. Bibliography: Description of the genealogical book. Miscellany: It's not categorized in the above four classification, including some unique documents (**Figure 1**).

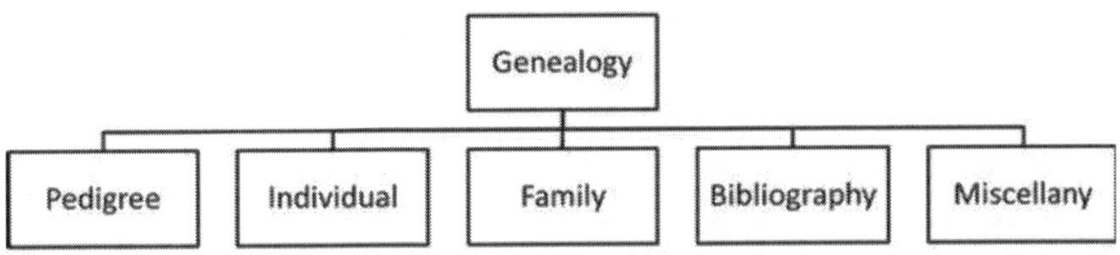

Figure 1. The sub-constructs of genealogy.

Pedigree consists of the following three constructs: Surname, First Name, Generation and Lang (Male). "Lang" is the digital name transformed from the legacy of the Central Plains culture, is unique characteristics of Hakka Genealogy (**Figure 2**).

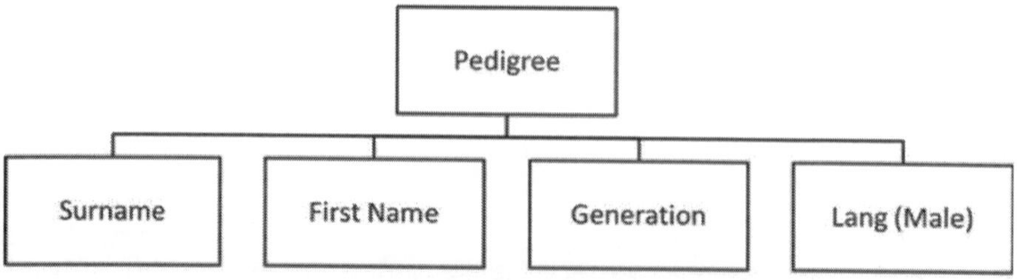

Figure 2. The sub-constructs of pedigree.

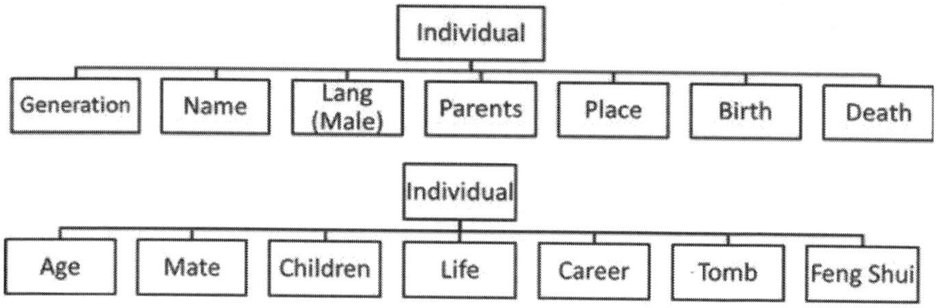

Figure 3. The sub-constructs of Individual.

Individual comprises the following constructs: Generation, Given Name, Lang (Male), Parents, Present Place, Date of Birth, Date of Death, Age, Mate, Children, Experience, Career. Tomb is divided into Tomb Place and Feng Shui. It is worth mentioning that Feng Shui is a unique culture of Hakka. Hakka people feel that Feng Shui has a mysterious power and will affect their fortune (**Figure 3**).

The history of Hakka family comprises the following constructs: Origins are divided into Origin of Surname, Native Place, Migration Process, The name of Ancestral Hall. Honor included files record that offspring to become the officer. Regulation contains the Family Instructions. Ancestral Hall contains Family Temple. Ancestors of Celebrities to be recorded Given Name, Birthplace, Portrait, and Experience. It's characteristics of the Central Plains culture that don't forget own origins. Therefore, Hakka people will carefully record own origins and ancestors. Hakka people comply with the teachings of own ancestors and build family temple. In addition, Hakka people pay attention to the glory. They hope family is proud of its own. Hakka people will record the glorious deeds of their ancestors, and regarded this as his glory (**Figure 4**).

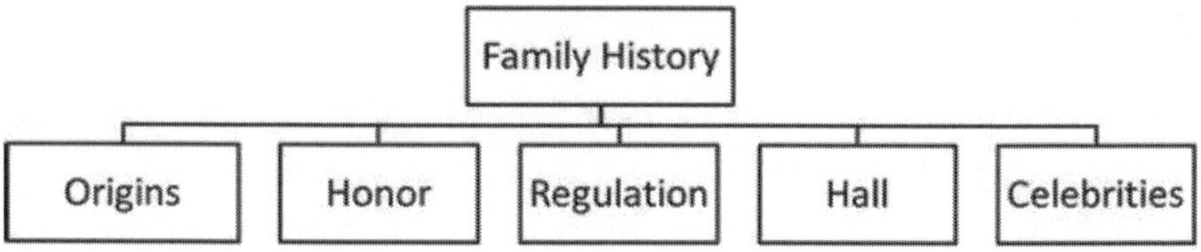

Figure 4. The sub-constructs of Family History.

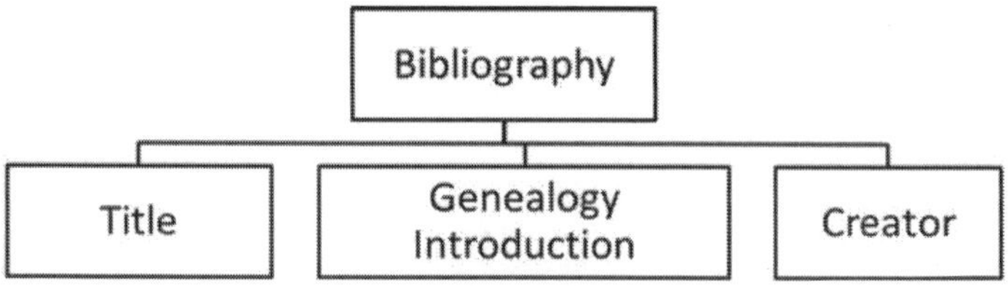

Figure 5. The sub-constructs of Bibliography.

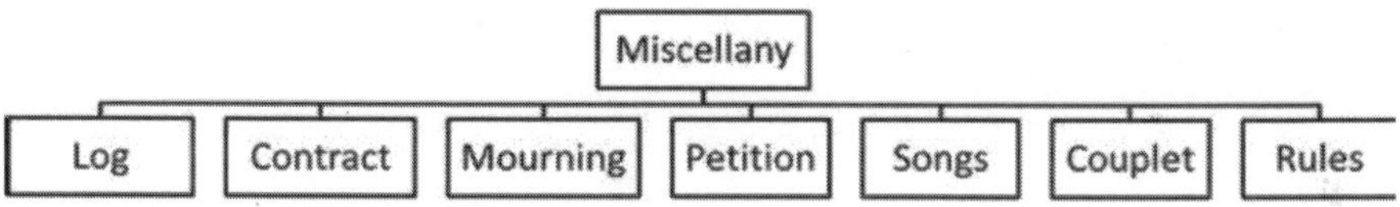

Figure 6. The sub-constructs of Miscellany.

Bibliography comprises the following constructs, Title, Genealogy Introduction and Creator (**Figure 5**).

Miscellany consists of various files. This study can't give a fixed form. In this work, we provide collection of document as a reference. Miscellany comprises the following constructs, Renovation tomb of the Log and Contract. The deceased was mourning article, Petition, Surname of Songs, Door couplet of the Ancestral Hall, and Ancestral Worship Rules. It will write on couple that the name of the Ancestral Hall implied meaning. The couple will be posted at the door (**Figure 6**).

The pedigree metadata standards for Hakka to implement the union catalog are presented in **Table 2**. The pedigree metadata have a total of five constructs and 31 sub-constructs.

4.2. Case Study: Tseng's Genealogy Design

This study selected one Hakka Family's genealogy as the case. After the complete content analysis, this study takes Tseng's genealogy to exhibit how this designed metadata has been applied in a real situation. This study used three dimensions to describe Tseng's genealogy metadata which contains the following data:

1) Pedigree consists of the following constructs: First Name (Yue-Yun), Last Name (Tseng), Father (MingYu), Children (Yue-Fun), Generation (22). Please see **Table 3**.

2) Individual consists of the following constructs: Generation (22), First Name (Yue-Yun), Lang (Hang Shao Three Lang), Parents Father (Jin-Sheng Tseng), Mother (Mi -Mei Yu), Present Place (Wugu Rd., Miaoli City, Miaoli County, Taiwan), Date of Birth (1934/7/25), Date of Death (1986/9/12), Age (53), Tomb Place (Jhonghua Rd., Miaoli City, Miaoli County, Taiwan), Feng Shui (West toward East), Career (Policeman), Mate (Si-Mei Liu), Children (Chang-Ren Tseng), Life (Born in Miaoli. He graduated from Miaoli Elementary School.). Please see **Table 4**.

3) Family History consists of the following constructs: Origins (Noble offspring), Native Place (Miaoli), Migration Process (Family migrated in 1785 from Taipei to Miaoli because the war), The name of Ancestral (Hall

Pengcheng Hall), Honor (Files record that offspring to become the officer Chin Tseng, General), Regulation Family Instructions (Failure is the mother of success). Please see **Table 5**.

Table 2. The pedigree metadata standards for Hakka to implement the union catalog.

Constructs	Items	Subitems
Pedigree	First Name	
	Generation	
Individual	Generation	
	First Name	
	Lang	
	Parents	Father
		Mother
	Present Place	
	Date of Birth	
	Date of Death	
	Age	
	Tomb	Tomb Place
		Feng Shui
	Career	
	Mate	
	Children	
	Life	
Family History	Origins	Origin of Surname
		Native Place
		Migration Process
	Honor	The name of Ancestral Hall
		Files record that offspring to become the officer
	Regulation	Family Instructions
	Ancestors of Celebrities	First Name
		Birthplace
		Portrait
		One's Life

Table 3. The Pedigree metadata for Tseng's genealogy.

Constructs	Items	Subitems	Value
Pedigree	First Name		Yue-Yun
	Last Name		Tseng
	Father		Jin-Sheng
	Children		Chang-Ren
	Generation		22

Table 4. The individual metadata for Tseng's genealogy.

Constructs	Items	Subitems	Value
Individual	Generation		22
	First Name		Yue-Yun
	Lang		Hang Shao Three Lang
	Parents	Father	Jin-Sheng Tseng
		Mother	Mi-Mei Yu
	Present Place		1, Lienda, Miaoli 36003, Taiwan
	Date of Birth		1934/7/25
	Date of Death		1986/9/12
	Age		53
	Tomb	Tomb Place	The Miaoli Shihtan Hiroshi Court
		Feng Shui	West toward East
	Career		Policeman
	Mate		Si-Mei Liu
	Children		Chang-Ren Tseng
	Life		Born in Miaoli. He graduated from Miaoli Elementary School.

Table 5. The Family History metadata for Tseng's genealogy.

Constructs	Items	Subitems	Value
Family History	Origins	Origin of Surname	Noble offspring
		Native Place	Miaoli
		Migration Process	Family migrated in 1785 from Taipei to Miaoli because the war.
		The name of Ancestral Hall	Pengcheng Hall
	Honor	Files record that offspring to become the officer	Chin Taeng General
	Regulation	Family Instructions	Failure is the mother of success
	Ancestors of Celebrities	First Name	Zi Tseng
		Birthplace	1956/1/1
		Portrait	
		One's Life	

5. DISCUSSION AND IMPLICATIONS

In this paper, we present designs the pedigree metadata standards for Hakka to implement the union catalog. Well-designed Hakka Pedigree Metadata makes it is easier to merge genealogy and provide offspring easy to find their ancestors. The Hakka Pedigree Metadata is comprised of many individual family and ancestors, each of which has a process in place for archiving and distribution of genealogy. Hakka genealogical metadata contains the following types of constructs. 1) Pedigree: use given name and lines expressed the genetically related. 2) Individual: describe the personal information. 3) Family History: the story of the ancestors is narrated. 4) Bibliography: description of the genealogical book. 5) Miscellany: included the unique documents that could not categorized in the above four classification. Results of this study can be applied to genealogies integration and implement the union catalog. In the future, information technology can be applied to genealogy integration and implement. The Hakka Pedigree Union Catalog is an attempt to make these individual family pedigree charts appear as one seamless genealogy to family and researchers seeking out the connection and family history.

REFERENCES

1. R. C. Griffiths and P. Marjoram, "Ancestral Inference from Samples of DNA Sequences with Recombination," Journal of Computational Biology, Vol. 3, No. 4, 1996, pp. 479-502.

2. A. J. Drummond, et al., "Estimating Mutation Parameters, Population History and Genealogy Simultaneously from Temporally Spaced Sequence Data," Genetics, Vol. 161, No. 3, 2002, pp. 1307-1320.

3. C. L. Liaw, "The Chinese genealogy," SMC Co., Ltd., Taipei, 2003.

4. J. M. Meskill, "The Chinese Genealogy as a Research Source," Family and Kinship in Chinese Society, Calif, 1970, pp. 139-161.

5. K. C. Zatloukal and N. J. A. Harvey, "Family Trees: An Ordered Dictionary with Optimal Congestion, Locality, Degree, and Search Time," Proceedings of the Fifteenth Annual ACM-SIAM Symposium on Discrete Algorithms, New Orleans, 11-14 January 2004, pp. 308-317.

6. M. K. Kuhner, "Coalescent Genealogy Samplers: Windows into Population History," Trends in Ecology & Evolution, Vol. 24, No. 2, 2009, pp. 86-93.

7. "Ancestry," 2012. http://www.ancestry.com/

8. National Central Library, "Taiwan Memory," 2012. http://memory.ncl.edu.tw/tm_cgi_test/hypage.cgi?HYPAGE=about_tm_e.hpg

9. C. Chen, et al., "The Design of Metadata for the Digital Museum Initiative in Taiwan," Online Information Review, Vol. 26, No. 5, 2002, pp. 295-306.

10. C. C. Yang and W. W. M. Chan, "Metadata Design for Chinese Medicine Digital Library Using XML," Proceedings of the 33rd Annual Hawaii International Conference on System Sciences, Maui, 4-7 January 2000, p. 10

11. Y.-N. Chen and S.-J. Chen, "An Analysis of Metadata Research and Application Development in LIS Based on a Hybrid Approach of Quantity and Quality," Bulletin of Library and Information Science, No. 55, 2005, pp. 51- 77.

12. L. Dempsey and R. Heery, "A Review of Metadata: A Survey of Current Resource Description Formats," The UK Office for Library and Information Networking, 1997.

13. DCMI, "Dublin Core Metadata Initiative," 2012. http://dublincore.org/documents/usageguide/

14. GEDCOM, 2012. http://www.gedcoms.com/

15. GedML, "GedML: Genealogical Data in XML," 2012. http://homepage.ntlworld.com/michael.h.kay/gedml/index.html

16. GEDC, "GEDC—An XML-Based Standard for Genealogy," 2012. http://www.sunflower.com/~billk/GEDC/

17. GEDCOMX, "GEDCOM X," 2012. http://www.gedcomx.org/Data-Models.html

18. J.-X. Xie, "The Hakkas Whose Roots Are in Central Plains," Journal of Huanghe S&T University, Vol. 9, No. 3, 2007, pp. 15-19.

19. X.-J. Li and X.-Y. Li, "On Hakka's Values from Family Rules and Instructions of Their Pedigree Records," Journal of Longyan University, Vol. 23, No. 5, 2005, pp. 71- 73.

20. S. Xie, "Hakka formation Development Outline of History," South China University of Technology Press, Guangzhou, 2001

21. X.-Y. Li, "Values Reflected in the Family Rules and Regulations of Hakka Genealogy," Journal of Guangxi University for Nationalities (Philosophy and Social Science Edition), Vol. 28, No. 3, 2006, pp. 77-80.

Chapter 4

How do Researchers in the Humanities Use Information Resources?

Ellen Collins, Michael Jubb

Research Information Network, UK

ABSTRACT

Information is at the heart of research. Every stage of the research cycle involves discovering, accessing, generating, manipulating, interpreting or presenting information, in order to advance knowledge. Researchers operate within a complex information environment, with needs that they themselves perhaps do not always fully understand, and are dependent upon services which are fast-changing and affected by much wider social trends. This paper examines how researchers in the humanities currently access, use and share information, paying particular attention to the influence of new technologies in changing information — and consequently wider research — practices. It finds that researchers are adopting new technologies where this helps them to work more effectively, but that traditional practices still dominate in some areas. There is mixed evidence of new research questions emerging as a result of engagement with new technologies.

Key Words: information resources; humanities; new technologies; research cycle

BACKGROUND

Several factors influence how researchers engage with information throughout the research cycle. These might be grouped into two broad categories. First, the supply factors — in other words, what information is made available to researchers, and how; and how are they able to reuse and disseminate that information? Second, the demand factors, and specifically the role of researchers' learned behaviours and accepted norms in determining how and where information should be found, analysed and shared. As we shall see, new technologies are particularly important influences upon both supply and demand factors, shaping the nature of information itself and the routes by which researchers access, analyse and disseminate it.

In terms of the peer-reviewed, published body of literature, the move to an online environment has been fairly well documented. Data for the United Kingdom show that higher education library expenditure on electronic-only serials has climbed from 9.6% of all serial expenditure in 2001 to 48.6% of such expenditure in 2009 (RIN, 2010). As of 2008, an average of 99.6% of all large publishers' titles were available online; even small publishers offered a very creditable 88% of their content in electronic format (Cox and Cox, 2010). UK researchers have seized upon this increase in provision, with article downloads more than doubling in the three academic years between 2003/04 and 2006/07 (Nicholas et al., 2009). As a result of this increase in provision and usage,Nicholas et al. (2011) suggest that researchers' habitual preference to 'power browse', or seek the key messages in articles rather than read the whole thing in a linear fashion, is being supported. In addition, the way that researchers locate such information is changing — the same research showed a strong preference for 'gateway providers' such as Google, PubMed or Web of Science, as opposed to publisher or library websites (Nicholas et al., 2011).

Research does not turn exclusively upon re-reading what others have written, of course, and the way that researchers handle data has also been affected by changing technologies. As Jim Gray recognised, technological advances have altered the way that researchers capture, curate, analyse and visualise data at every scale (Hey et al., 2009). This is particularly noticeable in the sciences, with cyberinfrastructure or e-science programmes in several countries dedicated to seizing the possibilities offered by large-scale computing. However, it is also beginning to apply in other fields, with disciplines such as digital humanities establishing themselves within universities, although they have not necessarily received the same level of support for developing their infrastructures (Borgman, 2009).

When it comes to sharing their research findings, academics are again presented with a widening range of options thanks to changes in technology. The democratising influence of the web has been felt in academia, as elsewhere, and there has been considerable interest in the role that social media can play in research communication. Researchers have indicated that such tools are particularly useful at the dissemination stage of research (Nicholas et al., 2010), but habitual usage is generally relatively low (Nicholas et al., 2010; Proctor et al., 2010). However, almost half of researchers make their work in progress available to their own network, and 37% make their data available online within their private network (Proctor et al., 2010). Publishing developments such as PLoS One, an open access online-only journal which publishes all articles judged to be technically sound, have also made it easier for researchers to share their findings, although the evidence suggests that publication in such journals is not yet esteemed as highly as more traditional resources (Proctor et al., 2010).

This illustrates the importance of the second set of factors which influence researcher behaviour — researchers' own habits. Although these habits are personal, there are discernable trends. Age and seniority of researchers can affect decisions to engage with information resources; for example, older, more senior researchers are more likely to use web 2.0 technologies (Proctor et al., 2010). Like other studies, this confounds the notion that the 'Google generation' of young researchers are somehow more at home with new technologies (Carpenter et al., 2010). Similar distinctions have been found in browsing habits, with more junior researchers less likely to browse. The authors speculate that this may be because they have less time to devote to maintaining an awareness of current research. They also highlight the importance of networks of collaborators and colleagues in raising awareness of significant contributions to the literature, and stress that these are built up over the course of a career, once again putting more established researchers in a different positions to those at the start of their working lives (Bennett and Buhler, 2010).

Discipline is widely recognised as another crucial influence on information behaviour (Tenopir et al., 2004; Nicholas et al., 2009). Even within a discipline, there can be identifiable sub-groupings: Tenopir et al. (2005) found that US medical faculty with MD degrees are less likely to read content online than medical faculty with PhDs. Nicholas et al. (2009) found that researchers in the same subject but different institutions use different strategies, facilities and functions when reading electronic journals. And at an even more local level, the habits of colleagues within a research group or collaboration can have an important effect upon information behaviours at every stage of the research cycle (RIN and NESTA, 2010).

Few researchers have attempted the herculean task of understanding all academics' information behaviours at every stage of the research cycle. A range of studies have looked at use of specific information tools or services — for example, web 2.0 (Nicholaset al., 2010; Proctor et al., 2010) or e-journals (Nicholas et al., 2010). Others have adopted a narrower focus, looking at how researchers within a discipline or institution use certain tools or services (Tenopir et al., 2005; Nicholas et al., 2009). But we still lack a detailed understanding of how researchers within a specific discipline create, use and manage information as an integral part of an entire research project, and of how such behaviours are changing.

To address this gap, the Research Information Network commissioned a series of in-depth case studies, looking at researcher behaviour in different disciplines. The first set, on researchers in the life sciences, was published in 2009 (Williams et al., 2009), and the third, on the physical sciences, will be published later this year. This paper presents the results from the second study, looking at researchers in the humanities (Bulger et al., 2011).

The traditional notion of the humanities researcher as a lone scholar in a dusty archive has long been redundant. Recent work found that 65% of humanities researchers had collaborated beyond their own department in the previous five years (Meyer et al., 2009). Other studies suggest that researchers are engaging with new technologies in the humanities, and are aware of how this may shape their research, in terms of both process and outputs (Warwick et al., 2008). This study attempts to understand how these developments in practice are changing the way researchers handle information within their research, in six different cases. It also looks at how the changing nature of information handling, and in particular the technological developments of recent years, has changed the nature of research itself.

METHOD

The cases were selected to offer a diversity of practice, including traditional methods alongside more advanced digital resource use. To ensure that the full range of humanities practices were represented, the researchers recruited participants through different kinds of cases, looking at users of a specific resource or database and researchers within a traditional department, within a field and within a collaborative project. The cases were as follows:

- Resource: Old Bailey Online, a database containing the criminal proceedings of London's central criminal court from 1674–1913.
- Resource: Digital Archive of Medieval Music (DIAMM), a database containing medieval polyphonic music manuscripts dating from around 800–1550.
- Department: University of Birmingham English Department, which has 45 members of academic staff and combines the study of language and literature.
- Department: UCL Philosophy Department, ranked first in the UK Research Assessment Exercise (RAE) in 2008 and with staff covering a range of research areas including philosophy of mind, history of mathematics, ethics and metaphysics.
- Field: Corpus Linguistics, a 'born-digital' field where researchers study a 'corpus' of texts from a variety of sources including newspapers, literature, spontaneous talk or broadcast.

- Collaborative Project: The Digital Republic of Letters, an international humanities collaboration developing an online research collaboratory of seventeenth-century correspondence.

Although the research team conducted a series of interviews and focus groups within each case to gather a range of experiences, the overall findings cannot be taken as representative of the humanities as a whole. There is probably a disproportionate representation of technologically-enabled researchers, as at least three of the cases involve some kind of digital resource. But, taken together, the cases show the diversity of practice and illustrate how researchers are engaging with new technologies, so that support services including funders, publishers and librarians can better meet their needs.

Findings

Search and retrieval strategies differed considerably between fields, depending upon the kinds of resources that are available. Historians using Old Bailey Online, for example, tended to use the database as their first port of call when seeking information, before turning to Google or other more generic services. Similarly, philosophers took advantage of the well-established Phil Papers pre-prints archive to find the most up-to-date publications. But there is still a strong awareness of resources provided by the library, which was specifically mentioned by users of Old Bailey Online, and by researchers in the philosophy and English departments. Members of the Digital Republic of Letters project mentioned the importance of direct contact with their colleagues in identifying relevant literature and data.

Researchers' approach to managing the information that they find is, in their own words, 'haphazard'. Researchers using Old Bailey Online created mini-databases of useful content for each of their projects, either on their hard drives or using online tools. Participants in the Digital Republic of Letters project cut and pasted important sections of their primary texts into Word documents, to be reordered and dropped into publications at a later stage. Many participants wanted better organisational tools linked to the data sources themselves, and where these were available they were well-used — DIAMM, for example, offers annotation and bookmarking facilities, which were popular with the musicologists.

As the literature suggests, collaboration is relatively common within the humanities, although not necessarily formally organised. Again, there are noticeable differences between subjects, with philosophers saying that collaborating to collect and analyse data was not encouraged in their field. Other disciplines were more enthusiastic, using the web to share their data reasonably openly among a community of interested peers. DIAMM was initially established as a repository for images of delicate or rare medieval manuscripts, and is therefore equipped with a facility for researchers to upload their own transcriptions. Many do so. In other fields, the infrastructure is not provided and researchers must find their own way to share. For example, corpus linguists use personal or discipline-specific websites to share their corpora and computer code that they have developed for analysis.

Perhaps because of the informal nature of collaborations, the way that humanities researchers communicate with each other also tends to be relatively informal. Most researchers, when asked about project communications, mentioned email, telephone and face-to-face meetings as their main channels. Even the most complex collaboration within the study — the Digital Republic of Letters — relied primarily upon these methods, with researchers flying internationally to share their expertise where necessary. Some researchers had begun to engage with web 2.0 tools for sharing work in progress, including Dropbox (Old Bailey Online), Skype (DIAMM) or wikis and Google Docs (English). But on the whole, traditional methods prevailed.

As the literature suggests, social media are viewed in a more positive light as a way to communicate research findings. There was some evidence of researchers blogging and tweeting, particularly younger researchers in philosophy and corpus linguistics. Younger philosophers mentioned that they used blogs to stay up to date with the newest contributions in their field. Philosophers also post papers on university websites prior to publication and receiving feedback, both solicited and unsolicited. However, this was not a new practice made possible by the web, but rather a continuation of a long tradition of pre-publication discussion, which previously happened at conferences, on email lists or face-to-face. This is just one example of researchers adopting new technologies where they enhance existing practices.

This pragmatic approach meant that in areas where new technologies do not meet every existing need there was a continuing demand for print; this was true of all the cases. Sometimes this resulted in an interesting hybrid approach. The philosophers, for example, used electronic versions of key texts to search for important words or passages, but used the print editions of the same texts to cross-reference page numbers. Academics in the English and philosophy departments made extensive use of Google Books to skim chapters or sections and decide whether it was worth acquiring the full text, either by buying the book or obtaining it via the library.

In other instances, the continuing demand for print seems to be based more on habit than objective judgements about usefulness. Most DIAMM users mentioned 'a notion in the field that you can always get more out of seeing the original than seeing a digital image of it', and a consequent pressure to use hard-copy manuscripts in archives, even when they themselves saw the benefits of using electronic versions. Similarly, some DIAMM users and researchers in the philosophy department felt that it was more prestigious to publish in print journals than in electronic-only ones, although they did not necessarily agree that this ought to be the case. This may be linked to the habit, which researchers in English, philosophy and the Digital Republic of Letters project admitted, of citing print versions of sources, even when the majority of the research has been conducted using an electronic resource.

One important question for the RIN project, and one which has been asked repeatedly in relation to e-science, is whether the new tools and technologies had fundamentally changed the nature of research. The answer, as with more scientific subjects, is complex. Researchers in most of the cases agreed that electronic resources enable new kinds of research, but do not necessarily inspire it. Old Bailey Online, for example, allowed researchers easily to search for every occurrence of an individual's name. But most interviewees suggested that this is something they had always wanted to do but which had previously been impossible, not least because nobody would fund the necessary months of work in the archives. This was a theme in the responses of other participants, particularly the corpus linguists (who also deal with occurrences of words or small phrases within large bodies of text). Incidentally, the technology sometimes brought its own problems: Old Bailey Online users sometimes had to search for a dozen different spellings or versions of a name to find every appearance of an individual within the records, since a computer cannot recognise the relationship between, for example, 'Dunckerley' and 'Tunikerly' in the same way as a human.

Researchers also suggested that electronic resources helped to put their research questions in a wider context. For users of DIAMM, this benefit was experienced on two fronts. First, the high-quality images allow more detailed examination of delicate manuscripts, permitting researchers to look very closely at the tiny, tell-tale habits which distinguish the anonymous scribes from each other. Second, once these connections have been made, the DIAMM database can record them, making links between different records which allows subsequent users to place their own scholarship in its place within the discipline. DIAMM users, like Old Bailey Online users, felt that linked digital archives would further enhance this facility.

Researchers attached to the Digital Republic of Letters project were the only ones within the study to suggest that their research questions had changed fundamentally as a result of the new technologies. The connections that they have built between various archives and the visualisation techniques that they have employed to 'see' these connections, have inspired new questions about relationships and knowledge exchange in the seventeenth century. It is perhaps relevant that this project involved the greatest degree of interaction and parity between technologists and humanists; both worked together to build the technology and therefore were able to create something that would really advance both disciplines. In this way, they may have avoided the limitations outlined by one of the developers working on Old Bailey Online, who felt that researchers were limiting themselves to doing research that the tool supported, rather than changing the tools to ask new kinds of question.

CONCLUSIONS

The humanists in this study were, overall, pragmatic users of new technologies. When they felt that a new tool would enhance their research practices, they would experiment with it, and often subsequently adopt it. A note of caution ought to be sounded about the enthusiasm for data sourcing and management tools such as Old Bailey Online and DIAMM, since participants were recruited due to their use of these services and thus constitute a purposely biased sample. The same is true, to a lesser extent, of the corpus linguists, whose discipline is effectively 'born digital' — only possible through technology. Nonetheless, the use of these tools is echoed by the philosophers' use of websites to share their unpublished papers. In all cases, existing research practices are made easier, quicker or more effective thanks to new technologies.

However, it was clear that researchers made clear assessments of the value of new technology and, where they did not consider it to be useful, would not engage with it for its own sake. This was particularly apparent when it came to social media such as Twitter and blogs. While researchers recognised the value of engaging with a wider audience through such means, they also felt that it could become a distraction from the main business of research. In other instances, new technologies were not seen as sufficiently good replacements for older practices. For example, despite some awareness of tools such as Google Docs and Dropbox, most researchers continued to use email lists to share and edit work in progress with their collaborators.

In other instances, rejection of new technologies was less easily explained, and there is clearly some residual habitual attachment to print. The experiences of DIAMM users illustrates this: they were under pressure from their colleagues to continue using original manuscripts even though digital versions provide new research opportunities. Some researchers' desire to publish in print as opposed to online-only journals may also reflect this problem, although it is also true that print journals tend to have higher impact factors, so it may be a rational decision on the part of researchers.

Evidence about the impact of these changing technologies on research questions and the nature of humanities research, is mixed. Researchers on the

Digital Republic of Letters project were certain that the technologies they developed had changed the nature of the research questions they asked. In other cases, notably Old Bailey Online, the developers expressed some frustration at researchers' passive use of the tools; they saw potential for changing the nature of research questions, but felt that the researchers themselves were only interested in doing the same type of work, but more efficiently.

These findings should be of interest to librarians on a number of fronts. Researchers in most cases were aware of the library and the services it offers, and named it as one of the resources they would turn to in seeking information. Most tended to use Google as a first point of call, but later moved onto more specialised resources, such as those provided by the library. Linking the library catalogue to Google searches may help increase the visibility, and thus use, of resources.

The hybrid print-electronic use observed in several of the cases is also interesting. The use of Google Books to skim publications for relevant content before making decisions about purchasing or requesting books via the library is potentially important. Encouraging this practice could help ensure that new acquisitions are actually needed, and not requested by researchers on a speculative basis. The ongoing interest in both print and electronic versions of content, and the tendency to cite the former even where the latter has been the primary source of information, suggest that a push towards e-books will need to be accompanied by considerable advocacy and support for researchers.

Finally, several researchers mentioned their concern that digitised versions of source material are often partial and unsustainable. Many are created through project funding, leaving them vulnerable in the longer term. It is rare for an entire collection to be digitised (due to financial constraints), and equally rare for data to be linked across archives or platforms. The report on which this paper is based recommended that library and information services 'could help by providing a structure that maintains digital tools and archives, and promotes development of innovative resources' (Bulger et al., 2011, p. 77). In the current straitened economic climate of most university libraries, this may seem like a low and somewhat risky priority. However, the enthusiasm with which some of these tools are being used suggests that they could be a valuable asset for library services in the longer term.

ACKNOWLEDGEMENTS

We would like to thank Monica Bulger, Eric T. Meyer, Grace de la Flor, Melissa Terras, Sally Wyatt, Marina Jirotka, Katherine Eccles and Christine Madsen, who undertook the research upon which this paper is based.

REFERENCES

1. Bennett, D. and A. Buhler (2010): 'Browsing of E-Journals by Engineering Faculty'.*Issues in Science and Technology Librarianship*, Spring 2010.

2. Borgman, C. (2009): 'The Digital Future is Now: A Call to Action for the Humanities'. *Digital Humanities Quarterly,* 3(4)

3. Bulger, M., E. Meyer, G. de la Flor, M. Terras, S. Wyatt, M. Jirotka, K. Eccles and C.Madsen (2011): '*Reinventing research? Information practices in the humanities*'. Research Information Network.

4. Carpenter, J., L. Wetheridge, N. Smith, M. Goodman and O. Strujivé (2010): *Researchers of Tomorrow; A three year (BL/JISC) study tracking the research behaviour of 'Generation Y' doctoral students. Annual Report*. Education for Change.

5. Cox, J. and L. Cox (2010): *E-only scholarly journals: overcoming the barriers*. Research Information Network.

6. Hey, T., S. Tansley and K. Tolle (2009): 'Jim Gray on eScience: A TransformedScientific Method'. In T. Hey, S. Tansley and K. Tolle (eds) *The Fourth Paradigm: Data-Intensive Scientific Discovery*. Microsoft Research.

7. Meyer, E., K. Eccles, M. Thelwall and C. Madsen (2009): *Final Report to JISC on the Usage and Impact Study of JISC-funded Phase 1 Digitisation Projects & the Toolkit for the Impact of Digitised Resources (TIDSR).*

8. Nicholas, D., I. Rowlands, P. Huntington, D. Clark and H. Jamali (2009): *E-journals: their use, value and impact*. Research Information Network.

9. Nicholas, D., I. Rowlands and D. Wamae (2010): 'Charleston Observatory 2010: Are social media impacting on research?'. At *XXX Annual Charleston Conference*, Thursday 4 November 2010 [Accessed 21 June 2011 from http://www.ucl.ac.uk/infostudies/research/ciber/Charleston-2010.pdf].

10. Nicholas, D., I. Rowlands, P. Williams, D. Brown and D. Clark (2011): *E-journals: their use value and impact final report*. Research Information Network.

11. Proctor, R., R. Williams, J. Stewart, M. Poschen, A. Voss, H. Snee, M. Asgari-Traghi, M. Chen, R. Combs, R. Elvey, B. Harries, A. Pechurina and H. Brown (2010): *If you build it, will they come? How researchers perceive and use web 2.0*. Research Information Network.

12. Research Information Network (2010): *Trends in the finances of UK higher education libraries, 1999–2009.*

13. Research Information Network and NESTA (2010): *Open to All? Case studies of openness in research.*

14. Tenopir, C., D. King and A. Bush (2004): 'Medical Faculty's Use of Print andElectronic Journals: Changes Over Time and Comparison with Other Scientists'.*Journal of the Medical Library Association*, 92(2), 233–241.

15. Tenopir, C., D. King, P. Boyce, M. Grayson and K. Paulson (2005): 'Relying on Electronic Journals: Reading Patterns of Astronomers'. *Journal of the American Society for Information Science and Technology,* 56(8), 786–802.

16. Warwick, C., M. Terras, P. Huntington and N. Pappa (2008): 'If You Build It Will They Come? The LARIAH Study: Quantifying the Use of Online Resources in the Arts and Humanities through Statistical Analysis of User Log Data'. *Literary and Linguistic Computing*, 23(1), 85–102.

17. Williams, R., G. Pryor, A. Bruce, S. Macdonald, W. Marsden, J. Calvert, M. Dozier and C. Neilson (2009): *Patterns of information use and exchange: case studies of researchers in the life sciences*. Research Information Network.

Chapter 5

Introducing Web Archives as a New Library Service: the Experience of the National Library of France

Sara Aubry

Web Archiving Project Manager, IT Department, National Library of France, Quai François Mauriac, 75706 Paris Cedex 13, France

ABSTRACT

The collections held by the National Library of France (BnF) are part of the national heritage and include nearly 31 million documents of all types (books, journals, manuscripts, photographs, maps, etc.). New collection challenges have been posed by the emergence of the Internet. Within an international framework, the BnF is developing policy guidelines, workflows and tools to harvest relevant and representative segments of the French part of the Internet and organise their preservation and access. The Web archives of the French national domain were developed as a new service, released as a new application and made available to the public in April 2008. Since then, strategies have been and continue to be developed to involve librarians and reach out end users. This article will discuss the BnF experiment and will focus specifically on four issues: * collection building: Web archives as a new and challenging collection, * resource discovery: access services and tools for end users, * usage: facts and figures, * involvement: strategies to build a librarian community and reach out end users.

Keywords: web archives, archiving websites, collection building, resource discovery, end users, usage, France

COLLECTION BUILDING: WEB ARCHIVES AS A NEW AND CHALLENGING COLLECTION

The Internet has taken on an important role to play in our daily lives: e-administration, e-learning, e-business, online publications, digital arts, blogs and new public spaces dedicated to discussion and chat, to name only a few. Many activities have moved partly or shifted completely to the Web and new ones have been created. With a constantly growing number of Internet users (about 35 million today in France) and an increasing number of French websites (1.7 million domains are registered under the .fr extension alone, called Top Level Domain or TLD),[1] it became crucial to look into this type of publication and communication medium, which is still quite new to a national library.

A website differs from other types of publications in many ways:

- It is not fixed on a particular format such as a piece of paper or a track on a disk, but is dependent on a rather complex network infrastructure. A website is not a single PDF file (like e-dissertations), it is not a single JPEG or a TIFF image (like images and photographs), but it is multifaceted: a web page is a compilation of many elements (texts, images, scripts, style sheets, audio or video files, etc.) that may come from many places (local file system, database, another distant website, etc.) and which are assembled together as each web user views the site with a browser. An analysis of the harvest of a sample of 2.9 million French websites in 2007 showed that there were about 1,600 different Internet media types.[2]
- A website does not have a beginning and it cannot be read to the end. It is an intellectual entity which can be seen differently from one user to the

other. It exists in relation with a network of other websites linked together by hyperlinks, a network of links which is even stronger than that of a scientific publication based on citations and bibliographic information.

- Websites are very numerous and the Web has no frontier. According to the last Netcraft survey, there are more than 206 millions websites.[3] Most of them are accessible from any country in the world and may be part of a national collection according to national, legal or institutional collection development policy guidelines.

- Web content is always in motion just like a stream: web pages may be updated up to several times a day (on the home pages of dailies such as Le Monde and Libérationthere are small banners that indicate the content was updated a few minutes ago).

- Websites and web content are also ephemeral: a web page may disappear at any time and for many reasons: voluntary or involuntary withdrawal by the webmaster, non renewal of the domain name, disk crash or network access problems with the host server, etc. Web content linked to a specific event, either predicted or unpredicted, is particularly at risk. On the occasion of a cooperative selection and harvest project of political websites during the 2007 French Presidential elections, the Library of Lyon found that 52% of 421 websites they selected were either totally or almost closed five months after the poll.[4]

LEGAL FRAMEWORK

As has been the case each time a new type of material of expression and creation was invented, including various new technologies as they appeared in France, the BnF first experimented and then adapted its organisation to harvest, preserve and give access to these born-digital publications. After books, engravings, music scores, photographs, posters, audiovisual and multimedia documents, the time has come to archive websites as well.

The French Heritage Law, or 'Code du patrimoine', now incorporates Title IV (articles L131-1 through L133-1) of the DADVSI law 2006-961[5] (DADVSI stands for Droit d'auteur et droits voisins dans la société de l'information, which is a French adaptation of Directive 2001/29/EC of the European Parliament and of the Council of 22 May 2001 on the harmonisation of certain aspects of copyright and related rights in the information society).

Officially published on August 3rd, 2006, this law:

- extends the scope of legal deposit to the Internet in the following terms: 'is also liable to legal deposit every sign, signal, writing, image, sound or messages of every kind communicated to the public by electronic channels' (clause 39). The law applies to all types of 'online electronic publications' constituting a set of signs, signals, images, sounds or any kind of message, as long as they are made available publicly on the Internet. Not only websites, but also newsletters and streaming media are thus included in this definition;
- defines how Web deposit responsibilities are to be shared between mandated institutions: INA (the national broadcasting institute which is responsible for preserving the audiovisual heritage of France) will collect sites related to audiovisual communications (mostly radio and TV) and BnF will collect all other sites; a forthcoming decree is to enforce both selection and access procedures;
- specifies collecting strategies: at the BnF, Internet legal deposit does not require permission from publishers and gives priority to bulk automatic harvesting: 'Mandated institutions may collect material from the Internet by using automatic techniques or by agreeing to specific deposit procedures with the producers'. The law also stipulates that no obstacle such as login, password or other form of access restriction may be used by publishers to restrict this process.

SCOPE

Although most of the public Web can be viewed by anyone in France, it is technically and legally impossible to archive the entire Web. BnF is mandated to harvest websites of the 'French national domain', that is:

- as a core, any website registered within the .fr TLD or any other similar TLD referring to the French administrative territory (for instance, .re for the French island of La Réunion);
- any website (possibly outside of .fr) whose producer is geographically based on the French territory (this can usually be checked on the website pages or using specific servers);
- any website (possibly outside of .fr) which can be proved to display content produced on French territory (this last criterium is more challenging to check but leaves room for interpretation and negotiation to the Library and Internet producers).

HARVEST TOOLS AND METHODS

Although we speak of a legal 'deposit', websites are in fact not deposited at the Library by publishers. Instead they are harvested by pieces of software called archiving crawler robots or simply crawlers. An archiving crawler works like the indexing crawlers of search engines. It is a programme that browses the Web in an automated manner according to a set of policies. Starting with a list of URLs, it saves each page identified by a URL, finds all the hyperlinks in the page (e.g. links to other pages, images, scripting or style instructions, videos, etc.), and adds them to the list of URLs to visit recursively.

Technical parameters shape the crawler identity and behaviour (scope, depth, speed, exclusion filters, etc.) but as web technologies are very complex and evolve very quickly, the crawler encounters many technical obstacles that keep it from harvesting all elements of a website or even of a webpage. Web archives are thus often incomplete. BnF is using the Heritrix open source crawler, developed in partnership with International Internet Preservation Consortium (IIPC) member institutions.[6]

Since it is not possible to aim for exhaustiveness or to undertake a manual selection of sites the BnF has chosen to combine two complementary harvesting methods in order to meet the challenges posed by Web legal deposit:

- bulk automatic harvesting of French websites: broad crawls intended to take a snapshot of a few thousand files from a very large number of websites. For instance, for the 2010 broad crawl, which is still running at the time this article is being written, BnF is harvesting a maximum of 10,000 URLs for each of 1.6 million websites. Crawlers harvest content without any distinction between academic, institutional, commercial or pornographic content. This method is truly in the tradition of legal deposit (i.e. it does not presume to identify what will be of interest to researchers in 100 years). However, archives that are created by this method are very superficial; they do not to keep track of deep content or website evolutions;

- focused crawls: selective harvesting complements the broad crawls. Subject librarians select websites for such focused crawls based on collection development plans (collaboration with other libraries and researchers is also possible). Focused crawls can be event-based (French elections in 2002, 2004, 2007, 2009) or thematic (personal diaries and blogs, sustainable development, Web activism, etc.). Focused crawls enable the building of more complete and more frequently harvested archives of a limited number of sites.

Today, the BnF Web archives or rather the 'Web archives of the French national domain' consist of 12.5 billion URLs and take up 145 Tb of disk space. The oldest web pages date back to 1996, and were acquired thanks to the Internet Archive, a non-profit organisation that aims to build an Internet library and is a pioneer in Web archiving. The most recent web pages date back to a few hours ago.

RESOURCE DISCOVERY: ACCESS SERVICES AND TOOLS FOR END USERS

Service Localisation and Access Restrictions

Giving access to these archives is not the same as giving access to documents physically located on the Library premises. Harvested websites are not recorded in the Library catalogue as the collection is too large and too heterogeneous; it would be impossible to establish an exhaustive list of archived websites, to know their titles and their detailed content. Instead, the BnF has built automated indexing processes to enable fast access to harvested content. Each file is dated and described to gather only necessary information (original location on the Web, format, size, localisation in the archives, etc.). This indexing process makes it possible to then replay archived websites within their publication environment and browse them by clicking links, just like on the living Web, but in a historical, dated context.

Since April 2008, Web archives are accessible to authorised users in the reading rooms of the Research Library, on the different locations of the BnF (Rez-de-jardin level at François-Mitterrand, and special collections departments at Richelieu, Louvois, Opéra, Arsenal, and Jean-Vilar in Avignon). Although the archives contain mainly publicly and freely accessible websites, this restriction was imposed to comply with legal provisions which apply to all legal deposit and heritage collections and which are intended to fully respect copyright and privacy regulations.

In order to be granted access, end users must be over 18 years old and give a proof of their need to access these archives for academic, professional or personal research activities (the BnF currently is the last and only resort for this type of document, since it is the only library to offer this service in France). Readers' cards are issued by the Readers' Guidance Service at the François-Mitterrand or Richelieu libraries after an individual admission interview with a librarian. Based on the users' needs this interview determines whether they are to be admitted to one or several departments and the card's period of validity (3 days, 15 days, annual).

The Web archives are available on 350 computers placed in all Research Library reading rooms, along with all the Library's e-services (information websites, booking facilities, catalogues, Internet access) and e-resources (digitised books and images, electronic journals, online databases, cd-roms, bookmarks). These computers are available to the public, but users need to make reservations for a place in the Library to be able to use them.

Searching and Viewing Tools

Web archives are accessible via a dedicated application named after the collection: 'Archives de l'Internet' (which means Internet Archives in French). The word 'archives' is used to emphasize that these collections are not exhaustive. The application is represented by an orange tin with the letters 'www' and a mouse pointer on it to show that although web pages are put in a box, they are still clickable. The BnF has chosen to develop a specific visual identity and a logo to make the Web archives more visible.

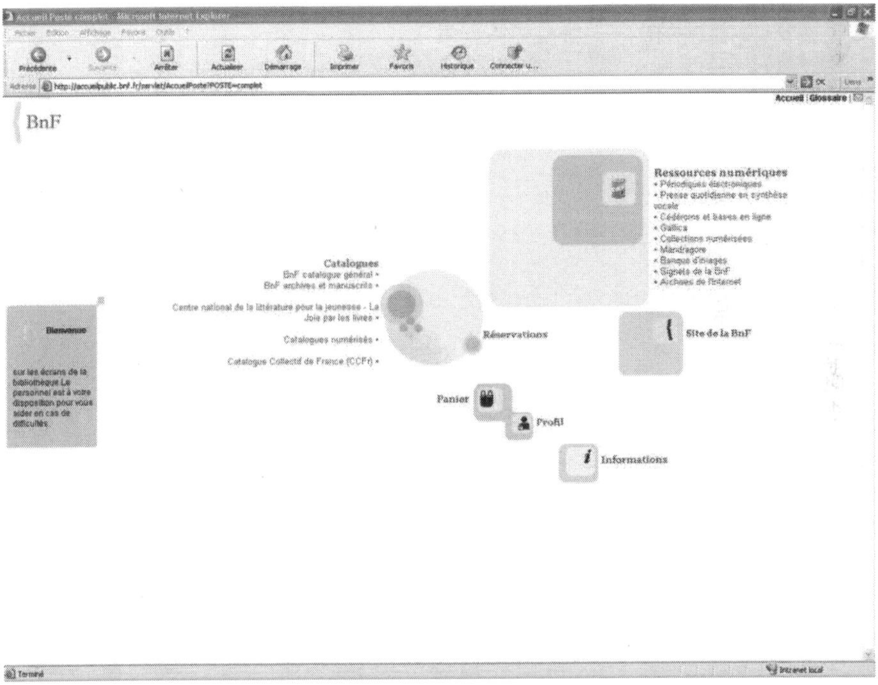

Figure. 1 Main screen of BnF e-services and e-resources in the Research Library.

To browse the archives, the BnF offers three different tools:

- search by URL,

- search by keyword,
- featured collections which aim to facilitate discovery of special collections.

All three tools have been integrated into a single application named after the content: 'Archives de l'Internet' (Figure 2).

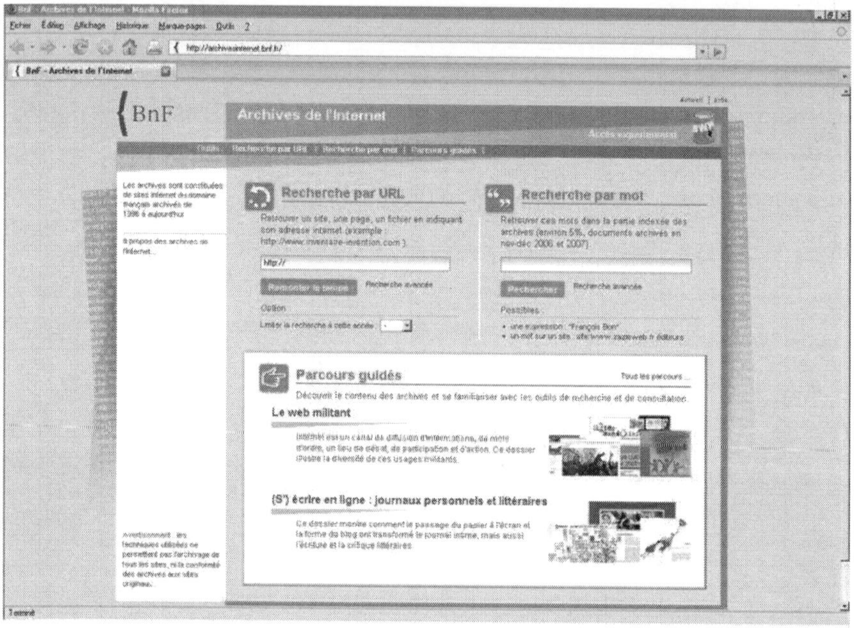

Figure. 2 Home page of the application 'Archives de l'Internet'.

- The URL search (URL stands for Uniform Resource Locator and refers to the Internet location of a web page) enables users to search the archive of a website, a web page or even a file by keying its original

  Internet location. For instance, searching http://www.lemonde.fr, gives 617 results (as of June 15, 2010) which are displayed in a calendar view from 1996 to 2010 (results can be narrowed to specific dates). Each date is clickable and gives access to the homepage of Le Monde as it was available at that date. This type of search is very useful when researching the evolution of a website and comparing versions, but just like for a call number search in a catalogue, users need to know the URL of the website or the web page they want to search (e.g. from a bibliography) or need to be able to find it (using directories or out-links referring to partner websites). Users may also browse the living Web and the Web archives in turns while using two visually different browsers simultaneously. Presently, the URL search is the only method to search the entire archives.

- The keyword search works like a traditional search engine: it enables users to search text documents containing a single or several words. Thanks to advanced search options, users may also search for an exact wording or a phrase or restrict the search to a specific website. This search is still experimental and only applies to 5% of the BnF Web archives. Full-text indexing of billions of unstructured files, taking into account duplicates and the temporal consistency that links them together is a technical challenge that several international research projects are trying to take on, but have not yet resolved.

Besides search options, the most important functionalities of this application are:

- The ability to display web pages collecting many elements that may have been archived at different dates, or at least at different times, and thus recreate a website artefact;
- The ability to click links taking temporal consistency into account (e.g. if we are looking at the French Presidency Elysée website on May 7, 2007 and click on a link to see the Government website, we expect to see the May 2007 version).

These functionalities and the URL search are supported by the open source Wayback Machine,[7] developed by the Internet Archive in partnership with IIPC. Keyword search is based upon the open source NutchWAX (Nutch with Web Archive eXtensions) software.[8] Also developed by the Internet Archive and by the Nordic National Libraries' forum, the latter is a tool for indexing and searching Web archives using the Nutch search engine and extensions for searching Web archives. To build an application based on these tools, the BnF has integrated them into its development tools and processes, customised them and extended minor functionalities (e.g. an implementation for persistent URL queries).

FEATURED COLLECTIONS

To compensate for the opaqueness of the collections and the lack of access tools (which are still evolving), but also to enlighten rich topical collections, BnF subject librarians in collaboration with researchers have built 'Parcours guidés' (literally: guided tours or featured collections in French), a set of illustrated and organised editorial pages that contain direct links to the Web archives, give an introduction, stimulate search and give ideas on other possible searches. Since 2008, three featured collections have been published:

- Click and vote: the electoral Web: a selection of websites from many types of producer (institutions, candidates, supporters, observers, individuals, etc.) from the 2002, 2004 and 2007 electoral campaigns (Figure 3);
- Writing oneself on the Web: personal and literary diaries: aims to look into the consequences of shifting from paper to the Web and the way blogs have changed personal, literary and critical writing;
- Web activism: shows how, over the years, the Web has been used by activists as a means of publication and communication, a lever to encourage commitment and a place to debate, participate, organise and act.
- As this paper is being written, three more featured collections are in preparation: sustainable development, amateur films and travel writing.
- As a complementary service, users are allowed to print web pages, copy and paste text samples in a note pad, print the note pad content or send it via email. Due to legal restrictions, there is presently no facility to enable users to take screen shots or grab web elements from the archives.

- Future development plans include full-text indexing and search as well as off-site access from the regional libraries sharing the legal deposit mission with the BnF around the country.

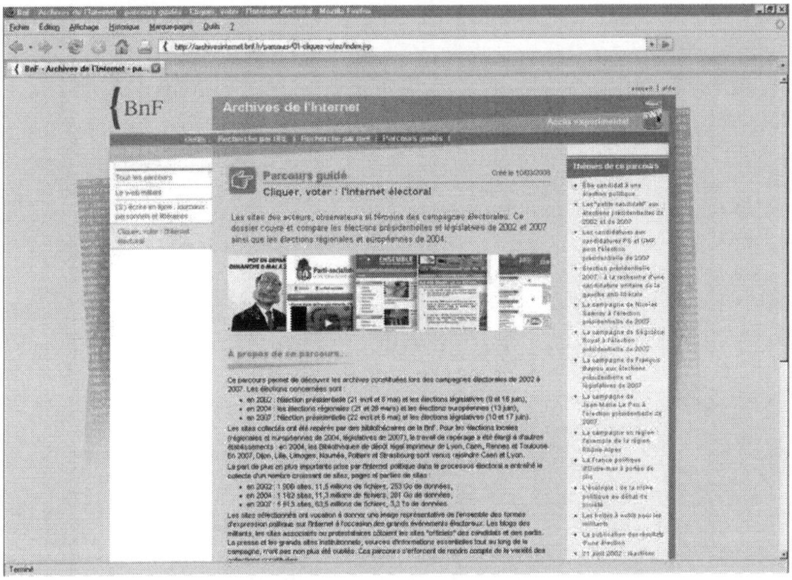

Figure. 3 'Click and vote: the electoral Web' featured collection.

Usage: Facts and Figures

The value of and the interest in Web archives will only be demonstrated over time, after web resources have disappeared from the Web; it is not part of the BnF's heritage philosophy and tradition to expect a short-term 'return-on-investment'; its mission is in keeping with a conception of the importance of time.

However, the evaluation of the first Web archives users and usage is an important step in setting up the Internet legal deposit. Demonstrating the public and scholarly usefulness of the harvested collections as well as developing and analysing the first usage statistics will enable us to measure the potential of these collections but also their limits. These analyses will be useful for both collection and tool development, in order to meet the expectations of researchers.

QUANTITATIVE ANALYSIS

An analysis tool called AWStats has been set up to analyse the real-time traffic on the application. Like the Wayback Machine and NutchWAX, it has been customised in-house to distinguish between public use by the readers, reference use by librarians at the admission or at the reference desk and professional use by BnF subject librarians.

Table 1. Main usage indicators in 2008 and 2009

	2008	2009
Average number of sessions per month	35	106
Total number of sessions	316	1275
Total number of viewed pages	35,891	90,063

Table 2. Detailed metrics 2009

	Visits	Visitors	Sessions (visits > 5 minutes)	Long sessions (visits > 1 h)	Viewed pages
Public	904	481	338	71	14,045
Reference	216	118	94	23	6242
Professional	1945	552	843	239	69,776
Total	3065	1235	1275	333	90,063

The two results that attracted attention at the BnF are that:

- the total number of sessions increased by three times between 2008 and 2009 (we will see later on that major training and information initiatives were developed for readers and reference librarians);
- there is a growing number of end users who are using the archives for in-depth research. The average session lasted 13 minutes in February 2009 and this increased to 30 minutes in December 2009. There were no more

than two public sessions lasting more than 'one hour or more' between February and September 2009. Then the number of such sessions increased to seven in October 2009, 32 in November, 22 in December and 23 in January 2010. These sessions indicate that extensive research is beginning to be performed on Web archives. Looking at the list of the most viewed pages, it looks like these projects are mainly being carried out by researchers in social and political sciences.

Qualitative Analysis

Complementary to the statistical analysis and a comprehensive list of the URLs and keywords searched, the BnF set up an internal, collaborative log used by librarians who met users on various occasions: admission interview, reference desk, personal meeting request, demonstration to websites producers, etc. Since April 2008, librarians registered 34 end users entering the following information: date/time, department and librarian name, request format (onsite, phone, email, ...), name, card type, research subject, notes/questions/comments/remarks (Figure 4).

Questions	Réponses
Date et heure	Mai / Juillet / août 2009
Dépt. ou service et nom de l'agent	Christine Genin (LLA)
Support de la demande	sur place
Prénom et nom du lecteur	
Titre d'accès	démonstrations
Sujet de la recherche/Discipline	- le 29 mai (17h30) : Elizabeth Legros, blogueuse et membre de l'équipe APA partenaire http://2009sediments.wordpress.com/ - le 20 juillet (16h) : Martine Sonnet, chercheuse au CNRS, écrivain et blogueuse www.martinesonnet.fr/blogwp/ - le 11 août (17h) : Silvère Mercier, qui a rédigé un billet dans son blog : http://www.bibliobsession.net/2009/03/17/archives-de-internet-demandez-votre-ticket-pour-la-posterite/ - le 21 août (17h) : Oriane Deseilligny, qui enseigne à Paris XIII, et qui est l'auteur d'une thèse sur les journaux intimes en ligne Elle s'intéressait surtout au parcours guidé et avait interrogé une collègue en banque de salle à ce sujet, laquelle l'avait renvoyée vers moi.
Observations	

Questions	Réponses
Date et heure	24 novembre 2009
Dépt. ou service et nom de l'agent	SOL Christophe Trébuil
Support de la demande	Sur place (Box d'accréditation)
Prénom et nom du lecteur	GREFFET Fabienne
Titre d'accès	Carte annuelle (délivrée le 24/11/2009)
Sujet de la recherche/Discipline	La campagne des élections européennes de 2009 sur internet
Observations	

Questions	Réponses
Date et heure	19/08/2009
Dépt. ou service et nom de l'agent	SOL
Support de la demande	Sur place
Prénom et nom du lecteur	DAFFINI Carole
Titre d'accès	Carte annuelle
Sujet de la recherche/Discipline	Les archives du Web (conservateur stagiaire)
Observations	

Figure. 4 Web archives end users log.

This simple tool enabled the BnF to identify those collections that are requested; for the moment these are in the area of social and political sciences.

On the day the BnF launched the service, a master student working on 'Internet and the 2007 Presidential election' asked for access permission. A Ph.D. linguistic student working on the analysis of women candidates' speeches came from Rome, Italy to work on communist leader Marie-George Buffet's blog which was closed just after the presidential poll. Nicolas Sarkozy's web campaign, the 'parti socialiste', the European 2009 elections, the use of videos in an election campaign, political cartoons and caricatures are a few examples of research in these areas.

Other searches have included such topics as: information searching in the European Union, archives of the Ministry of Ecology and its decentralised entities, amateur writers' websites, stress management websites, personal websites and questions about why and how websites are being harvested, and the rules of competition to help in a legal case.

Surveys

A first survey was held from October 2006 to June 2007 in the context of a master degree tutorial called 'Internet in campaign' and was co-organised by the BnF and an academic researcher in social media. 17 observation sessions and 5 interviews were conducted to identify users' needs in terms of tools and functionalities and their approach to a new media type (for their final exam, students had to write a paper on a topic which concerned living Web material and archived content).

A second survey is scheduled for November 2010. It will target current and potential users and is intended to define their needs in terms of collection content and identify means to develop usage (see Table 3).

Introducing Web archives in the Library means building a collective sense of ownership of this new type of collection both by staff and the end users.

The collections are in many ways cause for frustration (web captures are incomplete and quite a few websites do not appear in the archives at the moment). Similarly frustrating is the search application: for staff, in particular, the lack of a catalogue and the fact that librarians do not do any systematic descriptive work on the websites makes it hard for them to consider Web archives as valuable heritage collections. (How positive could a librarian possibly feel about a collection built by crawler software and not by him or her?)

Table 3 Framework for the next Web archive usage survey in November 2010

Current Web archive users	Potential Web archive users
5–8 interviews	5–8 interviews for each of the targeted groups: • scholars in other research areas (digital arts, humanities, media researchers interested in the Web itself, etc.), • a wider public interested in the memory of the Web, • lawyers and information professionals.
Questions about: • Collection development policy: selection, harvest periodicity and quality • Targeted users and usage types to develop • Interest in and themes for featured collections • Information and metadata given in displaying results and a page	Questions about: • Awareness about web archives • Interest identified in their own research area • Information types searched if needed • Actual usage constraints

Involvement: Strategies to Build a Librarian Community and Reach out to End Users

The challenge then is to highlight the distinctive qualities of Web archives — and notably the fact that they will be the only and unique testimony left of the major changes our society has been experiencing in the past 15 years and its transition from analogue to digital — but also to outline the patterns of this collection which actually makes it quite similar to older collections which librarians have learned to handle: apply the famous motto 'digital is not different' — and spread the word.

This being said, making web archiving business as usual primarily means using standard communication, organisation and marketing strategies.

Strategies to Make the Web Archives Part of Daily Library Business

Organisation. Managing the Web archives should not only be seen as merely a technical activity in the formal library organisation. Most organisations which have looked at web archiving as a primarily technical, IT-reserved topic requiring engineer and software development leadership have a hard time getting librarians involved in the promotion of the archives. In order to make web archiving business as usual, the BnF thus decided to implement its activities within the oldest production unit of BnF, the Legal Deposit Department. Websites are now managed together with legal deposit printed materials (books and periodicals) in a unit created in 2008 called the digital legal deposit unit. It consists of a team of five people, which runs the activities in partnership with the IT experts on the one hand and the collection experts on the other hand. This implementation within a big and long-standing department has helped a lot in including Web archiving experts and activities in the BnF's core mission and history. In other words, this type of organisation has helped making the new join the old, and demonstrating that large-scale sampling strategies for huge volumes of born-digital data were not so different from the way history has shaped the French tradition of legal deposit of printed materials over the past five centuries. This approach is part of a wider effort by the BnF to adjust staff organisation and skills to digital change.[9]

Networking. Building a network of subject collection experts all over the Library was another decisive step to encourage active contributions and acceptance of the Web archives as a collection by librarians and managers. In 2005, this network started as a group of 20 pioneers. By 2010 no fewer than than 80 librarians are somehow involved in selective harvesting, e.g. website selection, quality control and promoting resources and services to the public. Awareness was first brought to collection department managers. Most thematic departments consequently assigned subject librarians to the Web archiving project: in every major field (Arts, Literature, Philosophy, Sciences, etc.), including those involving rare collections (such as Maps or Music), there is now a 'correspondant du dépôt légal du Web' (a Web legal deposit correspondent), namely a person who has benefited from specific training and has acquired a basic knowledge of web archiving so that he or she can use it in their field of expertise. Many BnF subject librarians now acquire books and websites. From 2008 on, this network has been extended to more staff, 'les agents relais du dépôt légal du Web' (Web legal deposit shift agents), who were specifically targeted as reference librarians. The goal here was to develop knowledge of (and interest in) the Web archives in anyone involved with the public, whether in the reading rooms or online.

Tools for internal dissemination. The BnF is fortunate to be able to benefit from a rather important set of resources to promote information and skills internally: staff training, monthly internal staff talks and conferences at lunch time, a monthly printed internal magazine and of course the Intranet were all available before the start of the Web archiving programme. The challenge was to use all these communication resources and channel them in a smart way so that staff would hear about the Web archives at the right time and in the right way. For example, the BnF held a major internal presentation of the Election websites collection right at the end of the 2007 presidential campaign. Similarly, articles in the internal magazine or on the BnF Intranet follow a communication plan which takes into account the Library's general interest and trends. For example, when the Library launched a series of actions aimed at raising awareness about sustainable development (and to encourage staff to engage in responsible behaviour in this area), the Web archiving team chose to intensify its internal communications on collections related to sustainable development as well. In short, the internal communication plan to promote the archives was never seen as a stand-alone process. We take all possible opportunities to insert the Web archives in the Library's overall agenda rather than try to disturb this agenda by being or seeming somehow too intrusive.

Strategies to Reach the End Users

Controversies. 'Will the Web archives ever be used?' 'Why spend resources on this junk data while we have no guarantee that there is or will be public interest in this type of material?' 'Why not focus the efforts on the acquisition or digitisation of material that we know without a doubt to be heritage for the next generations instead?' — These rather typical questions often need to be answered, whether they come from the management or from the media. In a way, looking back at the history of the Library, such interrogations are not new either. Every new medium has raised similar questions over time. This is because it takes time — the time for things to disappear from the public, common trading space — before researchers or individuals realise they miss them, that they need this documentation to trace society's history.

All in all it is a typical situation for a heritage institution launching its Web archiving programme: it is urged to demonstrate tangible outcomes showing that it is useful and valuable, while at the same time usage cannot develop immediately. It is still too early for that, as most of the Web archives' users are yet to be born. The BnF's strategies to handle this situation have been two-fold: on the one hand, by providing tools and statistics which enable it to demonstrate the development of usage in a similar way as is expected for 'regular' collections (whether digital or not; see section above), on the other hand, engaging public debate to gain visibility outside the Library.

Create a public debate. Although not many people currently use the Web archives, most people actually show interest in this issue once they are asked. This is because the Internet has impacted everybody's private and public lives so everybody has something to say about it. The first reaction usually is: 'I never thought about it, but now you're telling me, I realise this is important. This is our memory and it is fading away everyday'. BnF communications target these people — those who might not show up at the Library to access the collections today, but who might develop an awareness that it is a job that must be done — in order to develop public support for the Library's Web archiving programme.

The 'Mémoires du Web' (Web memories in French) conference cycle was designed as a public forum to provide this type of support and visibility. It was launched in March 2010. Each conference lasts half a day and brings together three types of participants, which are also reflected in the composition of the audience: Web researchers, publishers and curators, all of whom are asked to deal with the same topic, e.g. Web activism and politics, Web diaries, the protection of personal data vs. the enlargement of the public space, etc. The BnF has brought together no fewer than 100 participants for each of these events and has received quite decent media coverage (mostly on the Web) for the first two sessions. The goal is to make Web archiving a subject of public debate outside of the Library, but together with the Library. This form of promotion (which also brings back excellent feedback for collection building and services to the researchers) complements more direct promotion activities organised for the current end users of the BnF.

Most users of Web archives do not exist yet. They will probably come with the forthcoming generations who are just born and will be born with the Web and who are using it or will use it as their principal means of information and communication. For a collection of that type, we must accept that only time will tell if we are making the right choice. Managing selection and access also means managing risks. We have to capture and promote what is unexpected and undesired today but that may be of interest tomorrow. We also have to envision tomorrow's users and usage. This issue is not new: it is a challenge any long-term heritage institution is familiar with. In the meantime, institutions still need mid-term strategies in order to build confidence and a sense of collective ownership of these new collections among their staff and stakeholders. Organisation, internal dissemination, communications and training efforts are key to develop new communities ready to 'adopt' the digital collection at large.

REFERENCES

1. Aubry, Sara (2008): 'Les archives de l'Internet: un nouveau service de la BnF',Documentaliste Sciences de l'Information, 45(4), pp. 12–13.

2. Bermès, Emmanuelle and Gildas Illien (2009): 'Metrics and Strategies for Web Heritage Management and Preservation', IFLA, http://www.ifla.org/files/hq/papers/ifla75/92-bermes-en.pdf.

3. Bermès, Emmanuelle and Louise Faudet (2009): 'The Human Face of Digital Preservation: Organizational and Staff Challenges, and Initiatives at the Bibliothèque nationale de France', iPres, http://www.escholarship.org/uc/cdl_ipres09.

4. Illien, Gildas (2008a): 'L'archivage d'Internet, un défi pour les décideurs et les bibliothécaires: scénarios d'organisation et

d'évaluation, l'expérience du consortium IIPC et de la BnF', IFLA, http://archive.ifla.org/IV/ifla74/papers/107-Illien-fr.pdf.

5. Illien, Gildas (2008b): 'Re-Inventing Collection Development Policy in the Age of Web Archiving: the Experience of the BnF', LIBER Annual Conference, http://www.ku.edu.tr/ku/images/LIBER/LIBER_ILLIEN_2008.ppt.

6. Lasfargues, France, Clément Oury and Bert Wendland (2008): 'Legal deposit of the French Web: harvesting strategies for a national domain', IWAW, http://iwaw.net/08/IWAW2008-Lasfargues.pdf.

Chapter 6

Library 2.0 and The Changing Landscape of Information Services in Academic Librariesin Kenya

Penninah S. Musangi

Senior Librarian – Karatina University, Kenya

ABSTRACT

Traditional models of libraries offering information services are not working for current generation of students. Thecurrent generation of library users can be termed as'digital natives'. These digital natives have grown up in an environment surrounded by rapidly evolving technologies. This means that today's academic library users think about technology and information differently than previous generations, and expect instant access to information. They can also be called 'netGen' users. With this kind of environment and expectations, academic libraries need to evolve by changing to offer user-oriented services and following users into their social spaces by allowing users to participate in the creation of content, keeping the users constantly updated and building services based on their feedback. To achieve this, libraries need to incorporate web 2.0 tools in the way they provide services. Some of the tools to take advantage of include creation of blogs, wikis, flickr, youtube, facebook, RSS feeds, socialbookmarks, microblogs, mashups and slideshare among others. Another obvious change noted today and especially in Kenya is the development of university campuses all over the country. For most of these campuses, one will find small rooms as libraries; with minimal resources and a few staff members. This is a challenge and library 2.0 being a combination of both physical and virtual spaces and services can offer solution to such campuses. In addition, many universities are offering distance learning and evening programmes to cater for the working class people who want to advance their careers. Do they have to come all the way to visit the library to get information for their studies? No. Library 2.0 is their way out and therefore academic libraries have to invest in these tools, in addition to off-campus access to e-resources. This paper will discuss library 2.0 as the new model of information provision to library users and especially in the academic world. The library 2.0 tools and technologies will be discussed and how they can be applied in an academic library.

INTRODUCTION

S.R. Ranganathan's Fifth Principle of Library Science states that "the library is a growing organism" (Ranganathan, 2006). In today's libraries, this principle is clearly exemplified by the number of new forms of information sources and services that are being offered. The boundaries of libraries have broadened to accept these new sources and services into their daily operation. The resources of the libraries have grown from physical objects to virtual objects, as evidenced by change from card catalogues to online public access catalogues (OPACs), and cooperative cataloguing to social cataloguing providing an ability to comment, review and reuse due to application of Web 2.0 tools (Ram, 2011). Web 2.0 tools allow users to create, change, and publish dynamic content of all kinds. Since they create a two-way or read/write environment they are enabling librarians interact with the library users (Stephens and Collins, 2007). Casey and Savastinuk (2007) provide what they terms as the 'truth' in relation to changes being experienced in the libraries; that libraries are losing interest of their users, they no longer consistently offer the services users want and they are no longer the first place current and potential customers will look for information.

They advise that, for a library to keep its current and reach potential users, library 2.0 is one avenue. Most academic libraries in Kenya have hinted to this advice and have not been left behind in this transformation, hence adoption of web 2.0 tools in their service provision.

Keywords: Library 2.0, Web 2.0, academic libraries, library service models, information services.

LIBRARY SERVICE MODELS

Kwanya et al. (2011) state that library service models can beperceived as types of library services which are differentiated through unique characterization such as type of collection held, target users, type of library (academic, research, school or public) in which it is offered, special features of the services offered, service philosophy, and general library organization. For a long time, academic libraries have offered information services using the traditional model of libraries being repositories of information and librarians being passive intermediaries in its transfer and interpretation. Then they moved to what is termed as 'hybrid' library model. This is where the traditional model runs parallel with digital model meaning offering services and providing information sources in both print and e-formats. This is the most common scenario in most academic libraries in Kenya today. However, this is not working well with the present generation of students. Robinson(2008) describes the current generation of undergraduate and graduate students as being digital natives. He adds that, these digital natives have grown up in an environment surrounded by rapidly evolving technologies. Therefore this means that today's academic library users think about technology and information differently than previous generations, and expect instantaccess to information which can offered through a digital library model. A digital library is a collection of services that are closely modeled on the traditional print library, but the information and the tools used to find the same is in electronic format.Since most academic libraries in Kenya cannot afford adigital library model, they have opted to revamp their hybrid models with the inclusion of web 2.0 in information provision, so as to cater for these digital natives. This has given rise to library 2.0 model.

Library 2.0 model

Kwanya et al. (2011) note that, library 2.0 is a model of library service whichharnesses the power of emerging information and communication technologies to create a dynamic physical and/or virtual library platform which is defined and controlled by the users and librarians and which facilitates the delivery of a superior library experience for the users: anytime and anywhere. Therefore, library 2.0 can be defined as a set of innovative technologies and services that are integrated with the library, facilitating the use of library resources and services, allowing library users to participate and keeping the librarians updated in their field (Zanin-Yost, 2010). Library 2.0 is a service philosophy that is guiding libraries in their effort to win new users while, at the same time, acknowledging that the current service offerings are insufficient and inflexible. Therefore library 2.0 can be termed as finding new ways of involving patrons by letting them contribute comments, add tags, rate library items, and get involved in other interactive and collaborative activities of the library.

Casey (2006) notes that library 2.0 sees the reality of the current user-base and says "not goodenough, we can reach more people". It seeks to do this through a three-part approach: reaching out to new users, inviting customer participation, and relying on constant change. Much of this is made possible thanks to new technologies such as web 2.0. Notess (2006) explores the importance of integrating web 2.0 tools into mainstream library and information services as it supports, promotes and extends information services to patrons or user spaces.

With this new service model, the focus in most academic libraries is changing to user-oriented services, by allowing users to participate in the creation of content, keeping the user constantly updated and building services based on users' feedback. To achieve this new model, libraries are incorporating web 2.0 tools in the way they provide services to their users. Some of these tools include creation of blogs, wikis, flickr, youtube, cluster maps, Google page translator, Google earth, facebook, LinkedIn, RSS feeds, and slideshare among others.

APPLICATION OF WEB 2.0 TOOLS IN INFORMATION SERVICES

Web 2.0 encompasses several technologies and services. These are as discussed by Stephens and Collins (2007) and Casey and Sarastinuk (2007), with examples of how academic libraries in Kenya have been able to apply these technologies:

Really Simple Syndication (RSS)

This is a web feed used for syndicating content from the internet. It informs users of updates to the subscribed sites. Many web-browsers have built-in-feed readers or aggregators that easily add feeds to web page. In libraries, this has been applied to:

i. Announce the availability of new resources in a particular subject to a particularclientele.

ii. Promote events organized by the library to its users.

Integrating library services through RSS feeds. Users can be informed of any newservice a library is introducing.

Examples of RSS feed in academic libraries in Kenya; http://opac.library.strathmore.edu/cgi-bin/koha/opac-search.pl?idx=kw&q=financial%20management&count=50&sort_by=acqdate_dsc&format=rss2

http://maktaba.ku.ac.ke/cgi-bin/koha/opac-search.pl?idx=kw&q=finance&count=50&sort_by=acqdate_dsc&format=rss2.

Instant messaging (IM)

This is real-time communication between two or more people based on typed text, or images. It is a popularmode of communication because of its quick response time, ease of use and the ability of alibrarian to multitask. Various programs are available to offer this service viz; livezilla, meeboetc. IM has been successfully implemented at Strathmore University Library whereby it is being used for:

i. Providing virtual reference services, whereby online library users can ask questions, requests and get responses instantly.

ii. Checking availability of colleagues at desks of work, instead of walking to an office and miss a colleague thereby wasting time and energy;

iii. Conducting online meetings, this does not require a member to be within campus to havethe meeting. This can be done remotely.

Social networking

Social networks enable messaging, blogging, streaming media, and tagging. MySpace and FaceBook enable users/libraries to share themselves with one another. Del.icio.us enables users to share Web resources and Flickr enables the sharing of pictures. Myspace and facebook are thetwo popular social networking sites used in academic libraries as identified through web analysis. Myspace allows organizations to create their own profiles, pages and can be used by libraries to create web pages. Facebook allows individual librarians to create profiles (Figure 1).

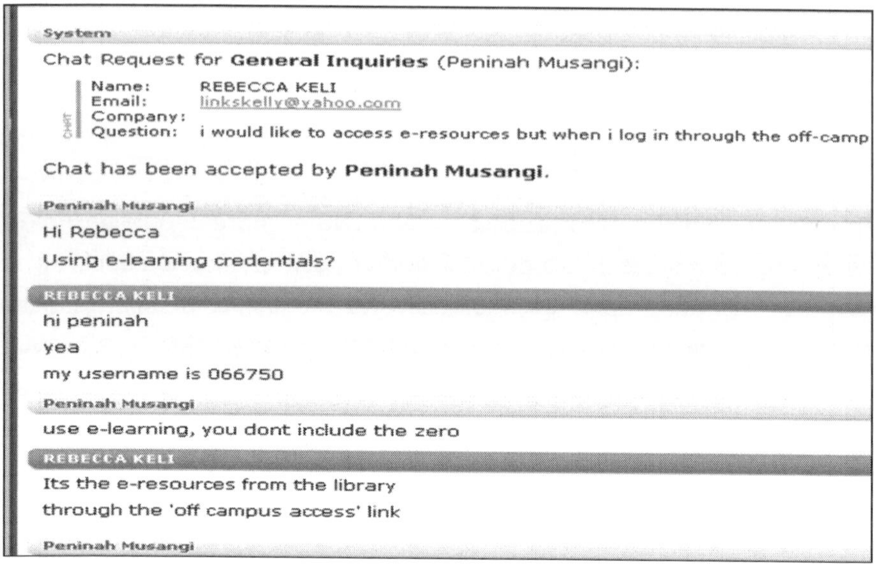

Figure 1. Instant Messaging.

This has been applied in:

i. Libraries creating a page to reach to new users. Most academic libraries have a Facebookpage, some with Twitter pages etc.

ii. Social networks enable librarians and patrons to interact, share and change resources. This can be achieved since users can comment or suggest what resources are useful.

iii. Building network among groups to discussissues of common interest. This is through the discussion forum in the social networks which give an opportunity for groups to discuss issues affecting them.

iv. Users adding content to the library catalogue. This includes book reviews or othercomments on particular resources in regard to their usefulness or applicability to specific subjects. Since there is no single text which can sufficiently cover a whole subject at university level, commenting on a resource on where it can be used in a particular subject is good and helps to direct other users.

Examples are: https://www.facebook.com/pages/Kenyatta-University-Library/239904162689026https://www.facebook.com/pages/Strathmore-University-Library/152833268081020

https://twitter.com/#!/strathlib_ke.

Tagging

A tag is a keyword added to a digital object to describe it. Tagging in a library essentially enables users to create keywords for the object at hand. As Maness (2006) describes, tagging is essentially Web 2.0 because it allows users to add and change not only content (data), but content describing content (metadata). Tagging has been extended to include social bookmarking where sites of interest, bibliographies or links to user-created content is collected and shared among the users. Academic libraries are using social bookmarking communities like delicious or Cite like. By using Flickr, users can tag pictures, while using Library Thing, they can tag books. In Library 2.0, users can tag the library's collection and thereby participate in the cataloging process and also tag items using terms relevant to such clientele. Tagging makes lateral searching easier.

Tagging has been applied in:

i. Library management systems for editing the subject headings or keywords to user point of view and thereby enhancing the indexing and relevancy of the searches, making the collection more dynamic.

ii. Facilitating lateral searching.

Blogs

Blogs are powerful two-way based tool, whereby libraryusers enter thoughts, ideas, suggestions and comments. A blog entry may contain text, images or links to other blogs and websites. A librarian or a library user can publish a blog post easily and cheaply through a web interface, and any reader can place a comment on a blog post.

Blogs have been used in libraries:

iii. To serve as a platform for users to file their concerns, queries and suggestions regarding the services and activities of the library.

iv. For collection development where users request resources.

v. As marketing tools of the information resources.

vi. As a tool for posting minutes of meetings for necessary actions.

vii. As a discussion forum.

Examples:

http://kemulib.blogspot.com/p/e-books.html

Stephen and Collins (2007) examine the principles of web 2.0, which have made it possible for library 2.0 to be conceptualized and put in place as discussed in the preceding section. These principles include:

i. Conversation: with the use of web 2.0 tools, user participation, discussion and feedback is possible and therefore encouraged.

ii. Community: the first principle of conversation can lead to a sense of community and belonging within the social sites.

iii. Participation: with the read/write capability, users create content, ideas, and knowledge which is shared between users leading to new information being created through collaboration.

iv. Experience: when users are engaged with others as a community, it is rewarding and leads to fulfilment.

v. Sharing: users are able to post as much information about what they know and enrich other users lives.

RECOMMENDATIONS

From the literature reviewed, libraries have to evolve in one way or another so as to provide services which are in tandem with the current generation of users, otherwise users will continue by-passing the library as the primary source of information. This can be achieved by incorporating web 2.0 tools among other technologies. In addition, the library needs to;

Provide information and instructionin as many formats as possible as students are accessing information using a variety of media such as video, podcasts and vodcasts rather than through the traditional channels of reading books or newspapers. Libraries should develop tutorials, videos and upload them on the library webpages or make use of YouTube, Slideshare. This can be done for orientation, guidesto e-resources, guide to use of Library Management System (LMS), information literacy.

Make library webpages and content "GYM-enabled" that is, optimized so that it is searchable by Google, Yahoo, Bing and other search engines. This way the library will continue to maintain its relevance to the general public. This can be done by installing 'add-ons' like LibX toolbar which enables users to search the library database like Google search.

Rather than simply have social networking sites like facebook, exploit on the everevolving features of the sites. For example, facebook in May 2007 launched a platform that allows users create applications. For instance a search features for a library catalog. This application is fairly simple for libraries to create since some of the users have willingly shared the code with others wanting to do something similar. Other features include creation of groups that facebook users can join, where messages can be send to all group members in order to advertise library programs, and services; ability to create a page that users can subscribe and get updates in their profile through news feeds. In addition to these features, the librarians can make use of the profile wall to answer questions, post items to discuss, albums to share pictures of library events and share scheduled events like workshops, training or library classes.

CONCLUSION

Today's library users want answers and want them now, and in the most convenient manner. Unfortunately the reality is that more than 90 percent of students whowish to find out about something, the first thing they would do, would be to search from the internet and not think about the library as an option. No one would think about asking a reference librarian or even logging onto a library website to use the inquiry service. This is the new marketplace; this is where most people live and work. Thus as libraries, we must recognize that the Internet and search engines are now the main ways in which people look for information. Therefore, rather than try to change users' habits, the library can change its approach and meet users where they are; on the Web, using the tools they enjoy using. To libraries, Web 2.0 may be nothing more than just new tools of technology, but how they are used and what they are used for will determine whether libraries can survive this new E-volution.

REFERENCES

1. Casey M (2006). Library Crunch: Born in the Biblioblogsphere. Retrieved August 6, 2012, from http://www.librarycrunch.com/2006/01/post_1.html

2. Casey M, Savastinuk L (2007). Library 2.0: A guide to participatory library services. New Jersey: Information Today.

3. Kwanya T, Stilwell C, Underwood PG (2011). Library 2.0 versus other library service models: A critical analysis. J. Libr. Inform. Sci. 44(3):145-162doi:10.1177/0961000611426443

4. Maness JM (2006). Library 2.0 Theory: Web 2.0 and its Implications for Libraries. text.Retrieved August 6, 2012, from http://www.webology.org/2006/v3n2/a25.html

5. Notess G (2006). The terrible twos: web 2.0, library 2.0, and more. Library, 30(3): 40.

6. Ram SKJPA,Kataria S (2011). Responding to user"s expectation in the library: innovative Web 2.0 applications at JUIT Library: A case study. Program: electroniclibrary and information systems, 45(4), 452-469. doi:10.1108/00330331111182120

7. Ranganathan SR (2006). The five laws of library science. New Delhi: Ess Ess Publication.

8. Robinson M (2008). Digital nature and digital nurture: libraries, learning and the digital native. Library Management, 29(1/2): 67-76. doi:10.1108/01435120810844658

9. Stephens M, Collins M (2007). Web 2.0, library 2.0, and the hyperlinked library. SerialsReview, 33(4): 253-256.

10. Zanin-Yost A (2010). Library 2.0: Blogs, Wikis, and RSS to Serve the Library. Library Philosophy and Practice (e-journal).Retrievedfrom http://digitalcommons.unl.edu/libphilprac/430

Chapter 7

The Challenges facing Professional Education and Practice of Academic Librarianship in the HE Sector in the UK

Emilce Sandra Rees

ABSTRACT

This article examines the challenges facing professional education and practice of academic librarianship in a changing Higher Education world. Librarians in the UK, like all the professionals in the academic sector, are faced with changes in organisational mission within the university, and the changing role of information within society. Inter alia, they have to be accountable inside and outside the university and are in competition for resource allocation and funding. Against this socio-political background, the following eight challenges have been isolated: 1. recognition of librarianship as a profession (and the accreditation of library and information studies); 2. ethical professional responsibilities in a changing organisational world, and how they can be weaved into the librarian's professional education; 3. the librarian in a changing (HE) world; 4. the "assault" on professionalism; 5. the user's centrality: fact or fiction?; 6. the librarian in an era of global challenge: is resistance futile?; 7. Potemkin villages: fabrication and performance in the academy, and 8. feminisation of the HE professions and patterns of employment. 9. Lastly, I'll examine future scenarios for librarians and the implications of present practices, which affect all sectors.

INTRODUCTION

This article examines the challenges facing professional education and practice of academic librarianship in a changing Higher Education world. Librarians in the UK, like all the professionals in the academic sector, are faced with changes in organisational mission within the university, and the changing role of information within society. Inter alia, they have to be accountable inside and outside the university and are in competition for resource allocation and funding. Against this socio-political background, the following eight challenges have been isolated:

1. recognition of librarianship as a profession (and the accreditation of library and information studies);
2. ethical professional responsibilities in a changing organisational world, and how they can be weaved into the librarian's professional education;
3. the librarian in a changing (HE) world;
4. the "assault" on professionalism;
5. the user's centrality: fact or fiction?;
6. the librarian in an era of global challenge: is resistance futile?;
7. Potemkin villages: fabrication and performance in the academy, and
8. feminisation of the HE professions and patterns of employment.
9. Lastly, I'll examine future scenarios for librarians and the implications of present practices, which affect all sectors.

Recognition of librarianship as a profession

In order to gain professional recognition by practitioners at large and by CILIP(Chartered Institute of Library and Information Professionals), an academic librarian is subject to the following career passages: firstly, a 'placement' on a graduate trainee scheme after a first degree but before library school, which might not be the 'reflective practicum'. Guile (2005) envisages for "the development of professional artistry" but, more often than not, this is an opportunity for academic libraries to acquire highly qualified trainees to perform routine tasks for a very low salary. After this year, the average library trainee takes either a postgraduate Diploma or Master Degree at library school, usually accredited by CILIP. It is then expected that after two years in a professional post the librarian should want to apply for Chartership status which, in theory, guarantees professional recognition and has been described by CILIP as a "seal of approval", a landmark which determines the transfer from the theoretical to the practical. This 'micro-management' of the profession, and the need for independent years of 'training' is, however, counterproductive to its ambition to be of equal standing to the major liberal professions.

Traditionally, nobody enters librarianship to maximise personal prestige and income, but out of a sense of vocation. CILIP, the representative body for librarians, is not made up of what Freidson describes as "an elite of unusually successful and distinguished practitioners" but of lesser mortals (Freidson, 2001, p. 142). However, 'a restriction of trade' via selection might be professionally welcome if the result is recognition and a rise in official status. For example, in most European countries, librarians have to pass selective national exams prior to library school, which guarantees the trainees a civil servant post for life.

Where is the librarian professionally located? Freidson includes the librarian in his classification of the professions needing a professional degree (Freidson, 1986, p.53). In Bourdieu's Homo Academicus, librarians are mentioned only once in his taxonomy of French intellectuals, under "Book Trade", in the company of publishers and booksellers (Bourdieu, 1984, p. 265). In its most elementary sense, 'professionalism' should be judged by the degree in which the members of an occupation can make a living while controlling their own work, with a degree of autonomy and of self-regulation (Freidson, 2001, p. 17).

A profession is not "mature" until it questions its role: recognition of professionalism should be achieved via true reflective practice, which should not be allowed to become a panacea or another professional buzzword. The concept of reflective practice was popularised by Donald Schon, whose work focused on the theory/practice problem of professional education in American Universities (Schon, 1987, Guile, 2005). Contra Schon, Barnett argues that Schon's conception of the professional is too individualistic, as professional life is necessarily social and intersubjective (Barnett, 1997). Guile is also critical of 'reflective practice', as it perpetuates the theory/practice divide (Guile, 2005). However, 'reflective practice' is repeated like a mantra by library managers (usually at training events) when at root level a culture of conformity is demanded and adhered to by employees, since, as Chomsky states, "if you show too much independence and question the code of your profession too often, you are likely to be weeded out of the system of privilege" (Chomsky, 2000, p. 24). The notion of 'reflective practice' might also subvert "dominant orders of discourse" (Morley, 2003, p. 115) while the quality 'experts' demand conformity.

Librarians must position their role within the academic community (the librarian as a scholar has still professional validity in Europe). If their role is interpreted as peripheral, this might lead to extinction: eight subject librarians were to be made redundant in a Welsh University. Information World Review (2005, March, p. 9) declared in their defence, that "the librarian is not a support role". As Morley wonders whether academics are being de-professionalised from scholars to knowledge workers (Morley, 2003, p. 91), we should in turn question whether knowledge workers have also sunk a notch down, to be considered mere office workers or highly trained clerical staff. In the post-modern economy, all jobs seem to have been reduced to a set of skills [1] - while at the same time segregation is encouraged. Neo-taylorian practices such as the "three e's" (economy, efficiency and effectiveness) have been adopted from the private sector by the present HE audit culture and new managerialism (audit itself being "a dividing practice" according to Shore and Wright, 2000, p. 61). Stephen Rowland remarks that the "separation between teachers and learners; the separation between academic staff and those who manage them; the split between teaching and research" is a deliberate action to threaten coherence (Rowland, 2002, p. 52). Fragmentation of knowledge workers from the rest of the HE professionals comes as a result of the ambivalent recognition of librarianship as a true profession by the rest of the academy.

Ethical professional responsibilities in a changing organisational world

In the public perception of what constitutes 'professionalism', ethical codes and trust are closely intertwined, as well as the assumption that the professional's self-interest comes second. Codes of ethics should be used to "create and sustain trust" and should serve to condemn "all those actions and circumstances in which the privileged position of practitioners is employed to generate profit beyond the value of the work that is performed" (Freidson, 2001, p. 215).

There is a range of factors that are seen as moral and ethical in each profession, which are consistent with philosophical definitions. CILIP has a responsibility to 'the public good' beyond the immediate limits of its membership. In the university, on the whole in the 'public sphere' (as conceptualised by Habermas, 1989), we find a commitment to rational debate and freedom of enquiry, an ethos which should also rule the professionalism of librarians. The CILIP code of professional practice lists the responsibilities that librarians have chiefly as employees[2], which are expected to be coterminous with those of the employers'. However, trust, rather than accountability, is argued to be the mainstay of professional practice (O'Neill, 2002), as current methods of performance accountability correlate with low trust (Avis, 2003), encouraging arbitrary and unprofessional choices.

The challenge to HE professionals (librarians included) hinges on not becoming accomplices in the administration of injustice, as professionals became in the Third Reich - where Berufsethos [professional ethos] was lost (Freidson, 2001, p. 131) and an unquestioning acceptance of the Nazi regime was shown by the universities (Barnett, 2000, p. 25). However, formal codes of ethics should not be mere manuals of administrative procedure, but should provide guidance for professional ethical issues - their disregard being, according to Freidson, another manifestation of an assault to professionalism (Freidson, 2001, p.216/17).

The librarian in a changing (he) world

A cursory glance at specialised HE library management literature shows that libraries are unwilling to innovate in any way that differs from the manner their academic peers are developing. "University A and University B [3] have taken different paths but arrived at the same destination" Moore states (Moore, 2003, p.151). Norry presents a case study, based on a University in the north of England, which identifies the following changes in staff roles, representative of the current trends affecting all universities - particularly those in the post-1992 sector:

- A tendency to conflate several roles in the role of 'librarian' (from different professions: ICT, information/library and technical media roles).
- Convergence of Library and IT departments.
- A desire for a strengthened partnership and liaison with academic staff - in terms of their pedagogic agenda, teaching delivery and learning models (**Norry, 2003**).

Interestingly, American HE libraries are preoccupied by the same set of issues that affect UK HE libraries: recruitment, education, and retention of librarians; the library as a place of intellectual stimulation and not a marginal service; impact of information technology on library services; preservation of digital resources; a change in the traditional library/publisher relationships; support of new users and reductions in HE funding, which impacts on library programmes, salaries, and resources, as well as the challenge of competition from other organisational units within the university (Hisle, 2002, p. 1-3). Some librarians, such as Brindley, Director of the British Library, while aware of these external and internal motives of change, are critical of the librarian's role as it has been constructed, and calls for diversity and dynamism in order to face these new challenges:

"Increasingly librarians will be judged as part of multi-skilled teams, as effective collaborators outside the comfortable box [...]. People, including librarians, are needed with higher level of skills and specialisms, and who are individuals willing to take risks, work in teams, be entrepreneurial, outgoing, manage projects, work quickly and approximately, and live in the messy world of the digital, where structured processes and procedures are often irrelevant. Not enough young people creating digital libraries see library school on their career path at present, or indeed libraries as somewhere exciting to work. Librarians have already missed many opportunities for engagement" (Brindley, 2001, p. 6).

In terms of defending our intellectual role, librarians as a group might not possess genuine monopoly over a body of specialized knowledge that could produce a 'knowledge gap', but individual librarians do, which they can use with considerable leverage over the user (Reeves, 1980, p. 137). If no professional librarians are used as information workers (in Freidson's terms, "if credentialism is not mandatory") library users must be trusted to make the right choices by themselves, a position that overestimates their discerning capabilities as consumers, especially in countries such as the US and the UK, where there is a high functional illiteracy rate (Freidson, 2001, p. 205).

The "assault" on professionalism

The professional issues addressed in this essay are not new. In the 3rd Annual Report of the CUKT dating from 31st January 1917 (alas, not a trendy French fashion label but the Carnegie United Kingdom Trust, which underwrote the activities of the Library Association) the following statement is found, and from which one can gauge how little the profession has changed its ways:

"The term 'librarian' is lightly used, and often applied to an official who is placed in charge of a collection of books, with very meagre knowledge of their contents and still less knowledge of the profession to which he purports to belong. This attitude on the part of some authorities results in a salary being paid which is totally inadequate for the responsibilities of the post. The inevitable result is that the profession as it stands today offers little or no attraction to a person with ambition, and who has a future to provide for." [quoted in Munford, 1976, p. 98].

In a new university in the north of England, most library staff roles were redesigned as part of the restructuring process in a review of academic services undertaken in 1996: "The most significant role change was that of Assistant Librarian to Information Adviser, demonstrating the deliberate move towards learner-centred support and hybridization" (Norry, 2003, p. 61). It is interesting how this author talks about the 'subject librarian' as an extinct species, and feels there is a need to explain what they were! (they are still 'alive and kicking' in Oxbridge, Russell Group universities, and in most so-called 'red brick' universities - as per D. Cook's classification of universities): "These were staff on academic grades who delivered teaching in information skills, were responsible for the acquisition and exploitation of resources in a subject area, and liaised with a specific school of study on information services and curriculum design and delivery" (Cook, 2005; Norry, 2003, p. 63). Retitling Librarians as Information Specialists, claims the author, sends a message of their information emphasis. In fact, what it sends a message of, is that their specialism is not respected anymore. The demotion of subject librarian to 'information specialist' is eclipsed by the importance given to the 'library managers', which are now in the limelight in HE in general: "Managers' roles were seen as vital to successful implementation of the change process and to making the new services work in practice" (Norry, 2003, p. 64).

Some regard specialisation in the library environment as counter-productive and undesirable. Smith refers to specialism[4] as a 'bad habit' picked up at library school and also as an internal unhealthy habit: "When professionals specialize to such a degree that they cannot understand each others roles, duties, or even terminologies, the effectiveness of the whole is diminished. [...] Academic librarians are typically more susceptible to the impairments of specialization than are public, school, or special librarians. [...]. Perhaps worse, library programs may recommend or even require that their students specialize in areas of librarianship. By actively limiting students in this way, programs inadvertently downgrade skills, undermining professional range and initial marketability. [...]A library staff composed of longtime, entrenched specialists also risks the development of territorialism [...]. Academic library patrons being at risk of poor service from a specialized environment." (Smith, 2003, p. 168-70).

Multi-tasking is favoured, as both the librarian and the manager of IT services are advised to form staff teams "to encourage multi-tasking and improve resilience to staff loss" (Buttle, 2004, p. 16). The argument is that qualified librarians can be 'saved' for more managerial positions (to the detriment of the reader's needs, as he/she no longer has access to qualified staff). The implication of this desirable 'multi-tasking' is that management do not want specialist librarians to mark their professional territory. Perhaps Freidson provides the ideology behind this anti-specialism position: "Since tasks and positions in firms are subject to change as productive ends and means change, commitment by workers to any particular job and body of knowledge and skill is obstructive and therefore undesirable" (Freidson, 2001, p. 109). If the professoriat is being proletarianized as a body by casualisation, part-time contracts, lack of tenure and other contemporary shifts (Readings, 1996, p. 1), this process is even more pronounced in the academic-related sector.

The user's centrality: fact or fiction?

The centrality of the library for students (as book prices soar) is rightly gauged by Barnett, despite his finding that the ritual [of visiting the library] is a strange affair, akin to a religious rite. "The significance that students attach to the library frequently emerges in student surveys: the opening hours and the availability of titles in both range and number (that is, duplicate copies) figure repeatedly as matters of concern to the students themselves. This pressure from the students is entirely proper." (Barnett, 1992, p. 39). In turn, students as customers of an institution should enjoy the same centrality. They have quite rightly become more demanding, seeking a return on their own financial investment in their education, with high expectations of the services and facilities provided. Reduced funding per student has put pressure on both schools of study and library services to reduce costs and increase efficiency.

Against this background, some academic library managers question whether libraries are really user-directed. It is ironic that Sykes, as Head of Library Services at the London School of Economics, which has an outstanding Social Sciences library, should posit this tentative worry. "Where collection development and the custodial role of libraries still tended to hold sway right into the nineteen-eighties, now the users stand much more at centre stage" (Sykes, 2002, p. 2). The implication of the centrality of the user is a change in the role of librarians, who now identify themselves with a new species of 'hybrid learner-support professional', rather than a librarian or IT professional. Tellingly, this hybrid librarian "sees more relevance in achieving membership of the ILT than chartership of CILIP" (Moore, 2003, p. 132) which is a questionable result, as it erases professional identity. The desirable solution resides in the retention of our professional identity as librarians, without having to mutate into technicians or teachers. It should be remembered that an approach such as Total Quality Management (in which all employees at all levels concentrate on satisfying the customer) was tried and abandoned by corporations (Shore & Wright, 2000, p. 69), and rated in the USA as too expensive, bureaucratic and unreliable (Morley, 2004, p. 10). Hence the centrality of the user must be tempered by the demands of operational practice.

The librarian in an era of global challenge: is resistance futile?

How information and knowledge are accessed in a global era makes the idea of the library as solely serving a local academic community parochial, as Universities see themselves these days as 'world players'. For Deem the globalised world impacts on the local scene, as HE must search for new sources of finance (e.g. commercial research) and accept intervention in the form of audits, through government agencies such as the QAA or the Quality Assurance Agency (Deem, 2001).

If the university is in ruins[5], so is the library as a scholarly model. The seminar library idea (each collection mirroring the faculty structure, and serving as a teaching unit) is becoming obsolete, but it's preferable to a non-discrete sequence of books as it is mostly the case in the libraries of the new universities. The library is not immune to international influences: the Humboldtian ideal of the seminar library was imported from Germany in the early 1800s, when Prussia was a Mecca for students, as America is today. However, the university as a repository of national culture no longer exists, and has been replaced by the 'University of Excellence' - an adjective that has no intrinsic meaning. With globalisation (or Americanisation?) comes the decline of the nation-state, where the university becomes "a transnational bureaucratic corporation" (Readings, 1996, p. 3). Some call for the resistance of the "McDonaldization of the world" (Barnett, 2000, p. 103), and not just of the university or the library.

Historically, there has always been 'global challenges' that have affected the university as an institution and have had a knock-on effect on the library. Political changes in Germany in the early 1990's after the fall of the Berlin wall required a 'purge' of university staff who were seen as part of the regime in the former East Germany. The university personnel appointed to replace them (who might have been of equal competence) were "less easily identifiable as ideological agents of their state" (Readings, 1996, p. 41). The same 'unification' was carried out in 1992 between the libraries of the (East Berlin) State Library and its counterpart in West Berlin (built anew in the post-war period). A senior British Library Briton was appointed in 2002 (after 10 years of an uneasy partnership between the libraries) to introduce the brave new world of library reorganisation and 'new management'. He only lasted in his post one year, leaving "over difficulties of bringing together staff divided between two workplaces" and "over problems with organisational regulations that effectively prevented dismissals and reassignment of posts" (Burchard, 2003). It seems that in this case, the moves from an outsider (who, among other things, was thought not to understand the relationships between East and West, and wanted to turn the East German library into a museum) were resisted successfully, thus it may be a case that Resistance (in tandem with efficient labour law) is not futile.

Some authors glimpse other possibilities: Barnett sees the new post-modern university as an identity where multiple standards, multiple purposes, multiple knowledges and multiple consumers intersect: "a distributed university, taking up the agendas and discourses of its many clients, regionally and nationally and probably internationally" (Barnett, 2000, p. 21). This is a scenario, which can only but benefit the heterogeneous composition of the academic library, in terms of services offered and users' composition.

Potemkin villages: fabrication and performance in the academy

Higher Education is not immune to fabrications or a display of "performativity" which is in evidence particularly during the Research Assessment Exercise (RAE). Putting on an act when subjected to the gaze of authorities is not a new stratagem. Grigori Potemkin (1739-91) erected fake villages along Catherine of Russia's Crimean inspection tour, while local peasants were given new clothes and ordered to look cheerful. These "Potemkin tours" of factories were also popular during the Soviet era to impress visitors and government inspectors. Individuals are complicit in the construction and maintenance of fabrications, as they succumb to what Lyotard calls the "terrors of performativity" (Lyotard, 1984). One of Louise Morley's academic interviewees (a senior lecturer) declares (Morley, 2003, p. 165): "We've got better at playing games in the research assessment exercise [...]. And in the subject review, people have got much better at hiding things". Performativity seems therefore to produce opacity rather than transparency and accountability (Ball, 2003, p. 215). 'To keep systems volatile, slippery and opaque' is also singled out as an undeclared policy of UK Higher Education systems of accountability"(Shore & Wright, 1999, p. 569).

There is even a new language to describe roles and relationships. In Morley's study, one interviewee, a head of an education department who became an assessor for subject review, describes his experience (Morley, 2003, p. 115) as:

"Well, basically you had to learn to write what was called QUAAHILI - QAA prose ... it was basically inducting people into a set of behaviours which were acceptable within the sort of defined norms of QAA"

This linguistic behaviour is described as a form of ventriloquism, as the compliance culture of present HE produces "ventriloquizing apparatchiks" (Morley, 2003, p. 162).

Self-monitoring (via self appraisal systems, target-setting, output comparisons) is also set into motion, without a concomitant right to reply. This lack of accountability on the part of reviewers reminds some researchers of the one-way gaze of pornographic objectification (Morley, 2003, p. 157). Explicit coercion is not needed, as this "audit society" has internalised hopelessness, or a TINA effect - there is no alternative (Morley, 2003, p.165). Some authors claim that there is a psychic operation at work, which "requires the activation and exploitation of a range of feelings such as guilt, loyalty, desire, greed, shame, anxiety and responsibilisation, in the service of effectiveness and point-scoring" (Morley, 2004, p. 4). Foucault describes the process of normalization brought in by "disciplinary power" as the introduction of "the constraint of a conformity that must be achieved"(Foucault, 1979, p.182/3).

The implication for professional practice is that professionals are all players, to a great extent, in this panoptical game, concerned with watching and being watched, from the "innocent" open plan office layouts to sophisticated "dataveillance" of military systems (Levin et al, 2002). The library is not immune to being an auditable unit like the rest of the university. The language of "best practice" and of "forever improving" is not alien in this environment, however empty these concepts might be once analysed in our daily professional practice.

Feminisation of the he professions and patterns of employment

It has been accurately remarked that credentials and academic capital, once attained by women, mysteriously lose their value. Figures from the Higher Education Statistics Agency (HESA) show that the average female in academia will earn four to five years' less salary than an average male colleague for the same number of years worked. "Quite often my perception was that permanent jobs were going to men, and usually public school educated men who were similar to people making the decisions. There was no pressure to show criteria, or to explain the decision, or any of these things" (Female academic interviewed in Morley, 2003, p. 153).

Some view the RAE as beneficial for women, as the new mechanism offers more opportunities for advancement, away from the "old boy" network: "Because [...] the more prestigious institutions are the ones with the most hidden snobbery, but they need to achieve good results in terms of the RAE like everybody else. If you've got a lot of research these days they tend to pick you even if you're not their type." (Morley, 2003, p. 153).

The shift from masculinised management approaches to feminised ones is judged to be a favourable one (Deem, 2001, p. 11). However, we must consider the implications of a feminised workforce in libraries, which might have contributed to the negative image of the profession and its lack of influence in a male HE world. A graphic portrayal of women in HE describes them as "organizational Morlocks - the creatures who live underground and service the 'Upper World' people" (after H.G. Wells's subterranean creatures in The Time Machine, quoted in Morley, 2003, p. 157). Traditionally, most head librarians are women, but they do not have the institutional power that male heads of departments have (in 15 years of personal professional involvement, I have not witnessed any of the head librarians in HE rise to pro-provost or vice-chancellors).

Casualisation is a pattern of employment, which affects academic-related staff and supports workers the most. Johnson-Eilola (1997, p. 79) muses ironically over the "flexibility" of casual employment: "a growing number of these workers are temporary, to be given an even greater degree of individual freedom at the end of the project: freedom from employment, from heath insurance, from paid vacation work or sick leave, or from any of the other benefits normally accruing to workers" (Johnson-Eilola, 1997, p. 79). What Eilola omits to state, is that this casualisation practice is feminised: women are 33 % more likely than men to be employed on fixed-term contracts (The Guardian,1999, quoted in Morley 2003, p. 153). The implications of casualisation will be examined in the next section, as it has repercussions for our professional futures.

Conclusion: for librarians, what futures?

While attempting to frame alternative courses of action, we must bear in mind the conservative nature of librarianship and the refusal to change (after all, the internet was not invented by librarians) which ultimately might see their fragmentation or mutation if not their demise in the 21st century. In terms of staffing practices, it is very likely that we are already witnessing Freidson's description of "a two-tier professional system composed of permanent, relatively small elite corps of professionals who do research and set standards of performance [...] and a floating population of qualified practitioners who may be employed on a temporary and sometimes part-time basis" (Freidson, 2001, p. 209). The librarians will have to get used to a "horizontal" as opposed to a "vertical" career (mobility from one organisation to another in the labour market, rather up the hierarchy). Supercomplexity as presented by Barnett may also constitute a viable form of survival for the library: the incorporation of multiple frameworks of understanding and of action, in an organisational world with "no unifying notions which served another age"(Barnett, 2000, p. 21).

Although the present situation of jobs' reductions is lamentable, many subject specialists are not actively promoting the value of their work (e.g. their engagement with the curriculum via user services and maintenance of academic collections), which put their jobs and everyone else's in jeopardy. Librarians must acquire the ability to bring what Barnett calls a "reflexive biography" to life, in order to face to the present challenges addressed here, and others to come (Barnett, 2000, p. 159). If we are to survive as a profession, librarians should shift from a responsive model to a pro-active model, by being at the vanguard of innovation, and not be seen as mere implementers or book selectors, as there is no time or place for conformity.

REFERENCES

1. Avis, J.: "Rethinking trust in a performative culture: the case of education". Journal of Education Policy, 18(2003)3, 315-332.

2. Ball, S.: "The teacher's soul and the terrors of performativity". Journal of Education Policy 18(2003)2, 215-228.

3. Barnett, R.: Improving Higher Education: Total Quality Care. Buckingham : The Society for Research into Higher Education and Open University Press, 1992

4. Barnett, R.: Higher Education: A Critical Business Buckingham : Open University Press, 1997.

5. Barnett, R.: Realizing the University in the age of supercomplexity. Buckingham : The Society for Research into Higher Education and Open University Press, 2000.

6. Blake, N., R. Smith R. and P. Standish: The Universities we need: Higher Education after Dearing. London : Kogan Page, 1998.

7. Bourdieu, P.: Homo Academicus. Cambridge : Polity Press, 1984.

8. Brindley, L.: "What use are librarians (working in libraries)?". Relay: the Journal of the University, College and Research Group, 51(2001), 5-7.

9. Burchard, A.: "Graham Jefcoate leaves the Berlin State Library just one year after taking up office - in conflict over the future of the institution". Der Tagesspiegel, Berlin, 13.3.2003.

10. Buttle, A.: Information Services Report, Number 895/04. Southern Universities Management Services, 2004.

11. Chomsky, N.: Chomsky on Miseducation. Edited and introduced by Donaldo Macedo. Lanham, Maryland : Rowman & Littlefield Publishers, 2000.

12. CILIP. When information goes wild... take cover or take control !, 2004 (Leaflet detailing the routes and qualifications for an information professional).

13. Cook, D.: Expansion and Innovation in British Higher Education, 1963-2003. Innovation and Change in Higher and Professional Education Module, 18th of January 2005, School of Educational Foundation and Policy Studies, Institute of Education, University of London, 2005.

14. Deem, R.: "Globalisation, New Managerialism, Academic Capitalism and Entrepreneurialism in Universities: Is the local dimension still important?" Comparative Education, 37(2001)1, 7-20.

15. Foucault, M.: Discipline and Punish. New York : Vintage, 1979.

16. Freidson, E.: Professional Powers: A study of the institutionalization of formal knowledge. Chicago and London : University of Chicago Press, 1986.

17. Freidson, E.:Professionalism: The Third Logic. Cambridge : Polity Press, 2001.

18. Guile, D.: We're all 'reflective practitioners now: a Utopian or utilitarian conception of HPE? Innovation and Change in Higher and Professional Education Module, 8th of March 2005, School of Educational Foundation and Policy Studies, Institute of Education, University of London, 2005.

19. Habermas, J.:The structural transformation of the public sphere: An inquiry into a category of bourgeois society. Translated by Thomas Burger with the assistance of Frederick Lawrence. Cambridge : Polity Press, 1989.

20. Hisle, W. L.: "Top issues facing academic libraries: A report of the Focus on the Future Task Force". C&RL News, 63(2002)10.http://www.ala.org/ala/acrl/acrlpubs/crlnews/backissues2002/novmonth/topissuesfacing.htm

21. Information World Review. March 2005, issue 211. Bangor wake-up call, p. 9.

22. Johnson-Eilola, J.:Nostalgic Angels: Rearticulating hypertext writing. Norwood, NJ: Ablex Publishing, 1997.

23. Levin, T., U. Forhne and P. Weibel (eds): CTRL [Space]: Rhetorics of Surveillance from Bentham to Big Brother. Cambridge, Mass. : The MIT Press, 2002.

24. Lowe, R.: "Higher education". In: A Century of Education, ed. R. Aldrich. London : Routledge Falmer, 2002, pp. 75-92.

25. Lyotard, J.-F.: The Postmodern Condition: a report on knowledge. Manchester: Manchester University Press, 1984.

26. Moberly, W.: Crisis in the University. London: SCM Press, 1949.

27. Moore, K.: "Building new partnerships: changing institutional relations". In: Centred on Learning: Academic Case Studies on Learning Centre Development, ed. E. Oyston. Aldershot : Ashgate Press, 2003.

28. Morley, L.: "Quality and Equality: Intersections and Collisions". In: Policies, politics and the future of lifelong learning, ed. A. Hodgson. London : Kogan Page, 2000, pp. 85-96.

29. Morley, L.: Subjected to Review: Engendering Quality in Higher Education. Journal of Education Policy, 16(2001)5, 465-478.

30. Morley, L. Quality and Power in Higher Education. Buckingham : Society for Research into Higher Education and Open University Press, 2003.

31. Morley, L.: Theorising Quality in Higher Education Bedford Way Paper 24. London : Institute of Education, 2004.

32. Munford, W. A.: A History of the Library Association: 1877-1977. Old Woking, Surrey : Gresham Press, 1976.

33. Norry, J.: "The changing staff experience". In: Centred on Learning: Academic Case Studies on Learning Centre Development, ed. E. Oyston. Aldershot : Ashgate Press, 2003.

34. O'Neill, O. Reith Lectures 2002. A Question of Trustt.http://www.bbc.co.uk/radio4/reith2002/

35. O'Reilly, D., L. Cunningham and L. Lester (eds.): Developing the Capable Practitioner.London : Kogan Page, 1999..

36. Oyston, E. (ed.): Centred on Learning: Academic Case Studies on Learning Centre Development. Aldershot : Ashgate Press, 2003.

37. Raish, M. (ed.): Musings, Meanderings and Monsters too: Essays on academic librarianship. Lanham, Maryland : The Scarecrow Press Inc., 2003.

38. Readings, W.: The University in Ruins. Cambridge, MA : Harvard University Press, 1996.

39. Reeves, W. J.:Librarians as professionals: The occupation's impact on library work arrangements. Lexington, Mass : Lexington Books, 1980.

40. Rowland S.: "Overcoming Fragmentation in Professional Life: The Challenge for Academic Development". Higher Education Quarterly, 56(2002)1, 52-64.

41. Schon, D.: Educating the Reflective Practitioner. London : Josey-Bass, 1987.

42. Shore, C. and Wright, S.: "Audit culture and anthropology: Neo-liberalism in British Higher Education". The Journal of the Royal Anthropological Institute, 5(1999)4, 557-575.

43. Shore, C. and Wright, S.: "Coercive accountability: the rise of audit culture in Higher Education". In: Audit Cultures: Anthropological Studies in Accountability, Ethics, and the Academy, ed. M. Strathern. London : Routledge, 2000.

44. Smith, A. and Webster, F.: The Postmodern University? Buckingham : The Open University Press & Society for Research in Higher Education, 1997.

45. Smith, K.: "On Specialisation". In: Musings, Meanderings and Monsters too: Essays on academic librarianship, ed. Raish, M. Lanham, Maryland : The Scarecrow Press Inc., 2003.

46. Sykes, J.: "Are academic libraries really user facing?". Relay: the Journal of the University, College and Research Group, (2002)54, 1-2.

47. Web sites referred to in the text

48. CILIP - Chartered Institute of Library and Information Professionals.http://www.cilip.org.uk/default.cilip

49. HESA - Higher Education Statistics Agency. http://www.hesa.ac.uk/

Chapter 8

Dealing with Maps: The Identification of Intellectual Responsibilities and the Application of Relator Codes

Sandra Domingues

Center for Geographical Studies, Institute of Geography and Spatial Planning, University of Lisbon

ABSTRACT

Treatment of information about those responsible for the intellectual content of work is one of the most complicated tasks in the bibliographic description of cartographic material. There are particular characteristics that make it difficult to apply some general rules laid down by technical cataloguing manuals for recording this information. The classic situation typified by the monographs — "an author and an editor" or "multiple authors and an editor" — differs from the large quantity of cartographic resources which have various contributors, with distinct functions, who contribute to the production of the resource. Field work, compilation of information and its standardization or homogenization, drawing, engraving or printing are some of the tasks often recorded on the maps associated with different individuals or organizations. In such cases, the problem is to select the person who should be considered as the principal author, determining with this choice the main entry of the record, and who should be chosen as co-authors or secondary authors. It is often difficult to know whether or not the functions performed by each one, or even by only one individual, deserve the epithet of "author" of the map.

In cartographic material a great variety of situations and many aspects can be taken into account, which allows evaluating the different solutions to problems that complicate documental treatment. Therefore, maps, loose or forming homogeneous groups, the nature of represented information, the time and institutional context in which they are produced and published, as well as whether they are printed or digital or manuscripts, influence the procedure for giving the authorial statement.

Modern maps (produced using digital tools), although they confront librarians with problems about the determination of access points, may be easier to treat than those created prior to the nineteenth century. However, this is not a rule: old maps are not always more difficult to treat than contemporary cartography.

Keywords: map cataloguing, authorship, relator codes

1. COMPLEXITY OF CARTOGRAPHIC PRODUCTION: AUTHORS AND FUNCTIONS

Cartographic production involves a diverse set of activities. The acquisition of spatial data through field surveys or using various information sources and various tools is one of the most important steps. In the case of topographic maps and until the advent of aerial photography, which occurred in the first quarter of the twentieth century, the work of data acquisition was exclusively provided by the classical method, through observations and measurements made in the field. After validation of the acquired data, all the items were harmonized and, finally, the map underwent a review process. If it was edited, there was also the final drawing and printing to be carried out. But this process, briefly presented, has of course, over time, shown great developments.

The decade of the 1960s represents an important milestone in the recent transformations of cartography. As result of the technological changes that began at that point, it moved "da produção manual (ou analógica) para a produção baseada em meios informáticos (ou digital) e, por consequência, dos mapas em papel para os mapas digitais", as well as "produção, antes quase exclusivamente centralizada em profissionais (cartógrafos) e instituições vocacionadas para as actividades cartográficas e hoje mais acessível a qualquer pessoa, munida de meios tecnológicos mínimos2" (**Dias, 2007**, p. 37). These changes altered cartographic production and were reflected in the statement of responsibility associated with cartographic resources, showing that the work of cataloguing should consider the context in which the resource was produced.

Before the nineteenth century in Europe there were two basic types of information sources: surveys made in the field or maps made by others. While surveying, carried out by surveyors and topographers or military and civilian engineers, demanded accurate theoretical and practical preparation, compilation from existing maps was also a difficult task, involving selecting and matching information from various cartographic sources held in the office. Some of the engineers, who were accustomed to taking measurements in the field using surveying instruments and to drawing terrain accurately to scale, developed into " géographes de cabinet", that is, compilers and even engravers or publishers of maps. They began to devote themselves to commercial cartography and the map trade. However, other map compilers did not have any training in field surveying.

Tomás López, geographer to the king of Spain in the late eighteenth century and compiler, editor and publisher, compared these two types of map production in his*Principios geográficos aplicados al uso de los mapas*: "El Geógrafo trabaja en su casa, teniendo a la vista papeles varios de un mismo terreno, que compara, y adapta lo que segundo su buena crítica es mas perfecto. No es ministerio suyo levantar planos particulares, porque para esto hay otra clase de gentes, que no necesita mayor instrucción, que la de llegar a saber hasta la Geometría rectilínea. Si los Geógrafos necessitaran ver y medir la tierra, que compreende sus Mapas, ninguno hubiera podido durante su vida publicar una de las quatro partes de la tierra3" (**Garcia & Moreira, 2008**, p. 122). "Assim, à semelhança de muitos outros geógrafos, cartógrafos e editores europeus, López elabora os seus mapas pelo somatório, confrontação e síntese de descrições geográficas e corográficas, e de mapas de escalas e tipos diversos, a denominada Cartografia de gabinete. A 'otra clase de gente' são os engenheiros, particularmente os militares, que representam os "modernos", e que defendem uma Cartografia com base no estabelecimento de triangulações e em levantamentos topográficos4" (ibid.).

Maps of that period often bear notes saying that they were based on field observations. These were intended to attest the truth of their authorship and their up-to-date content, because manuscript and printed maps could be copied exactly or with minor changes by any compiler or "géographe de cabinet". On the other hand, some map compilations skilfully combined literary sources and graphics from different time periods. The "géographe de cabinet" looked for commonalities or discrepancies between sources and synthesized similarities to produce homogeneous information. According to **Pedley (2005**, p. 25), the compilation is "the process of reading these literary accounts, taking various manuscript plans of the same place and coordinating them by determining their scale, analyzing the measurements used by the surveyor, and redesigning the maps on a projection that could accommodate the scale of the map". Until the nineteenth century, truly unique maps, resulting from costly field surveys or difficult compilation work in the office, might or might not be printed. This decision depended on several factors, including economic ones. Generally, the map scale is closely related to the intended purpose of the map, which also determined its dissemination. Large-scale maps were intended for use in official or public administration, often remaining as manuscripts. In contrast, medium and small-scale maps covering large territories were often directed at a broad audience, which was becoming more numerous over time, and were compiled by a "geographer" and reproduced by printing for commercial distribution.

In eighteenth-century Europe there were at least two types of division of labour in map compilation and engraving. In France, the compilers were normally not engravers (e.g. Delisle, Buache, or Gilles Robert de Vaugondy). In England, though, the compilers were often engravers, and even de facto authors, as stated by **Pedley (2005**, p. 45): "at least it is the engraver's name that often claims authorship of a map for lack of any other". The compiler status of Guillaume Delisle (1675–1726) is proven by information contained in the title of his own map of Europe (Figure 1). The map also reveals that the engraving work was not done by Delisle, who employed several engravers, each with a specific function (engraving the map borders, the water features, the topographical details, the lettering and the ornamental cartouche).

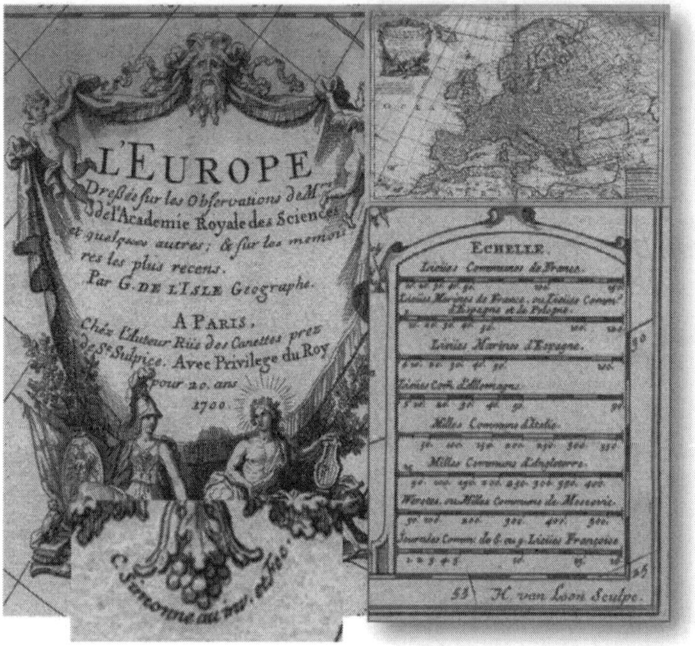

Figure. 1. Map of Europe compiled by Guillaume Delisle, dated 1700, and extracts showing the title and the area of graphic scales, indicating the various contributors: Royal Academy of Sciences, H. van Loon and C. Simonne (Source: National Library of Portugal).

It was the publisher (English) or "éditeur" (French) who played the leading role in the production and dissemination of maps and atlases to the general public for the simple reason that he was the owner of capital. The involvement of publisher in cartography was expressed in "contrats entre les éditeurs, les cartographes et les graveurs; des procès pour des plagiats ou du travail mal fait; des actes de vende de matériel d'impression, de planches gravées ou de fonds géographiques entiers" (**Pastoureau, 1980**, p. 452). In many cases, even though they had training as engravers, they would devote themselves exclusively to the commercial side of the business. Some of these map publishers even held the title "Ordinary Geographer of the King", incorrectly suggesting that they had responsibility for the intellectual work of map compilation (**Pastoureau, 1980**, p. 452).

This brief description of some aspects of map production before the nineteenth century illustrates the challenges faced by cataloguers trying to identify the person(s) responsible for the content of cartographic resources from earlier time periods. In summary, regarding the assignment by the cataloguer of primary responsibility for maps, it is possible to identify two categories in eighteenth-century European map production: those who produce maps from field surveys — topographers, surveyors, geographers and engineers — and those who start from written or graphic records of others — compilers or geographers of cabinet. Nowadays the generic designation of cartographers is attributed to both. Thus, whenever they are clearly identified, they should be considered as the main access point to the document. Among the remaining participants in map production, the draftsmen and engravers should be recorded with a secondary statement of responsibility by the map cataloguer. Publishers and printers should be confined to fields for publication and distribution; users searching the catalogue will be able to retrieve them through those access points.

In regard to old maps (**Andrew & Larsgaard, 1999**), the general recommendation is to transcribe into their bibliographic records not only all the statements of responsibility, but also all the other production and publication information that appears on them. As well as the difficulty of correct treatment of authorship when cataloguing old maps, there is the problem of hard-to-identify copying and counterfeiting of maps. The proliferation of counterfeit maps was due to the difficulty and expense of survey and compilation. Engravers, printers and booksellers, particularly during the seventeenth and eighteenth centuries, copied the maps of others or printed maps from engraved plates that they had acquired. Plagiarism was not even limited to the borders of the country, since foreign maps were easily copied. In order to make people believe that the map was an original compilation, indications were added to the statement of responsibility which referred to various credits, such as "amended and corrected from" or "improved from the map of".

To mask copies, printers, publishers and booksellers used the work of their engravers. They introduced decorative elements, fonts or symbols and even different scripts and dedications. The title of the map was often used to claim authorship and to clarify any doubt, by mentioning the names of all the people involved in its production, even with secondary responsibilities. The name of the person who ordered the work and who directed and executed the surveys helped to give even greater credibility to the work. In this way they affirmed that its preparation was based on surveys, removing the suspicions of forgery and making a new product using content which sometimes was more than a century old.

In the eighteenth century, printed maps were produced in a complex context: "The printed maps trace its source to a manuscript map that incorporated observations of space; these observations resulted from physical effort, such as a survey, or intellectual effort, such as compiled map, or a combination of both. The manuscript map may be copied, enlarged, or reduced, its information removed, altered, or expanded several times over before a version of it (or several different versions of it) is engraved on copperplate, ready for printing. Thus the information embodied by a printed map has already been distilled several times before engraving, a process that itself alters the content of the map" (**Pedley, 2005**, p. 198). Even after publication, the maps were still "copied and reprinted for different purposes, for different audiences, by different authorities, further transforming its contents, with different results for its users" (Ibid.).

In conclusion, for a cartographic resource there are many factors of responsibility, and we can categorize them further as follows: topographer, surveyor or whoever stood on the field in the case of topographic maps, hydrographic charts, etc.; compiler; draftsman; engraver; and, finally, the person responsible for a work in which the map is integrated (for example, maps inserted in an atlas or another publication).

2. IDENTIFICATION AND SERIATION OF AUTHORS IN BIBLIOGRAPHIC DESCRIPTIONS

Maps, like books, may be the result of individual or collective, personal or institutional work. In general, in the most recent cartographic resources individual authorship, even institutionally framed, is often explicitly mentioned. However, if it involves production organisations, authorship entered in the document refers almost exclusively to the organization and less often to the individual contributions of each contributor (though half a century ago this was not the general situation). Thus, in large part, the individual statement of authorship usually appears today associated only with thematic maps. In hydrographic and topographic cartography, on the other hand, the inscription on the documents generally refers only to the institution or official body responsible for producing the country information and keeping it up-to-date. The *Carta militar de Portugal 1:25,000* (1928 –), from the Instituto Geográfico do Exército, is a good example of the evolution in the expression of individual and corporate authorities: the first sheets published contain not only the producing institution, but they also reference some contributions of individuals: drafting, lithography, updating, etc.., which later disappeared (Figure 2).

Despite the large number of maps currently being produced by institutions, the references to individual responsibility cannot be ignored or eliminated. Some map librarians, including American ones, believe that: "when a personal author is given, it should be used as main entry" (**Rockwell, 1999**, p. 43), whereas others claim that also for maps produced by an institution, but containing expressions as 'charted by ….', 'designed by …' or 'compiled by …', the person linked to such an expression should be considered as the individual author and the main entry to the document.

The selection of the person who has the greatest responsibility for the work and who should head the catalogue record is not a simple task. The lack of clear rules and clarifying examples allows technicians to choose questionable procedures without uniformity. When there is an identified producing

	200	1	a	Caminha
			f	Serviços Cartográficos do Exército
			g	Trabalhos de campo Cap. Gaspar
			g	des. Cap. Varel
			g	des. litog.co João Marques
Edicao	205		a	[Ed. 1]
Cartograficos	206		a	Escala 1:25 000, projecção de Gauss
Publicacao	210		a	[Lisboa]
			c	S. C. E.
			d	[1953]
Desc. fisica	215		a	1 mapa topográfico
			c	color. ;
			d	40 x 64 cm
Coleccao	225	2	a	Carta Militar de Portugal 1:25 000.
			i	Continente, série M888
			v	fl. 14)
Nota Titulo	304		a	Estereofotogrametria aérea e restituição efectuadas pelos Serviços Cartográficos do Exército. Trabalhos de campo: 1949
Nota Edicao	305		a	Menção de edição e data atribuídas com base em informação do organismo responsável
Nota Publicacao	306		a	Impressão: Papelaria Fernandes
Nome Comum	606		a	Mapa topográfico
Geograf. Assunt	607		a	Caminha (Viana do Castelo, Portugal)
Geograf. Assunt	607		a	Viana do Castelo (Portugal)
CDU	675		a	912.469.11(084.3)
			v	BN
			z	por
Autor Secundari	702	0	a	Varela
Autor Secundari	702	1	a	Gaspar
Autor Secundari	702	1	a	Marques,
			b	João
Ent. Principal	710	01	a	Portugal.
			b	Serviços Cartográficos do Exército,
			f	1932-1959

Titulo	200	1	a	Caminha
			f	Instituto Geográfico do Exército
Edicao	205		a	Ed. 2
Cartograficos	206		a	Escala 1:25 000, projecção de Gauss, elipsóide internacional, datum de Lisboa
Publicacao	210		a	Lisboa
			c	I. G. E.
			d	1997
Desc. fisica	215		a	1 mapa topográfico
			c	color. ;
			d	40 x 64 cm
Coleccao	225	2	a	Carta Militar de Portugal 1:25 000.
			i	Continente, série M888
			v	fl. 14)
Nota Titulo	304		a	Levantamento, digitalização, processamento e desenho por meios automáticos executados pelo Instituto Geográfico do Exército. Estereofotogrametria aérea, trabalhos de campo: 1995
Nota Conteudo	327	0	a	Contém, impressa, a seguinte indicação: "Cobertura cartográfica do território espanhol baseada no Mapa Topográfico Nacional de España 1:25 000"
Nome Comum	606		a	Mapa topográfico
Geograf. Assunt	607		a	Caminha (Viana do Castelo, Portugal)
Geograf. Assunt	607		a	Viana do Castelo (Portugal)
CDU	675		a	912.469.11(084.3)
			v	BN
			z	por
Ent. Principal	710	01	a	Portugal.
			b	Instituto Geográfico do Exército
			f	1993

Figure. 2 Extracts from the bibliographic records of two editions of a sheet of the Carta Militar de Portugal 1:25,000, containing some explicit statements of authorship (extracted from: http://mapoteca.igot.ul.pt/).

institution and, in parallel, there is individual participation, there are some who use as a criterion for decision making the way in which the statements in the document are presented: if the largest contributions are from individuals, one of them should be the main entry of the register, with the institution being recorded as secondary (and vice versa). However, as is easily noted in the examples in Figure 2, this rule will rarely give the attribution of primary responsibility to the individual authors. The cartographic institutions fulfil today almost all functions simultaneously: the author (in his various tasks), the editor and even the printer or distributor.

Another example is provided by contemporary thematic maps, including geological or agricultural maps. In the first case, besides mentioning the public body for which the map is produced, the sheets still discriminate against those responsible for the field survey and the source of the cartographic base. This leads us to the question whether the authorship should not be given to the geologists who collected and interpreted the information so that the institution will only assume responsibility for editing. In these cases, according to some experts, there are only two possibilities: to consider the institution as principal author and assign a secondary entry to the geologist or vice versa. It is possible to give those that are responsible for base maps a secondary role, but the practice in many national and international map libraries is that they are only mentioned in the notes.

Regarding thematic cartography, it is also important to consider the effects those modern digital technologies for the construction of easily accessible and user friendly maps have had on the statements of responsibility associated with this type of map. The professionals provided with these tools and support on the Internet have become largely self-sufficient. Today, one professional can carry out the entire chain of production and, institutionally framed or not, will have full responsibility for the map produced. The responsibility of editing remains with the institutions for which they work, or with which they collaborate. The producers of basic geographic information and institutions that provide additional data for the preparation of such maps will continue to be assigned to the function of information sources, mentioned or not in the field of notes according to the documental policy established by each library.

The existence of multiple indications of authorship, associated or not with different functions in a given cartographic resource, leads inevitably to doubts about which of them is considered the principal entry. It is not easy to establish general rules, as it is hard to find stable criteria that support decision making about the various functions of the cartographic production chain, the enormous diversity of situations over the centuries and the nature and complexity of this information: "each map or map set being cataloged must be taken on a case by case" (**Rockwell, 1999**, p. 43).

3. CONSTRAINTS AND SOLUTIONS IN THE APPLICATION OF RELATOR CODES

The previous discussion about the clarification of the contributors in the complex production chain of cartographic resources leads inevitably to the question of attribution of relator codes to the different contributors responsible for the content of the maps, using the table published in the *UNIMARC manual*: bibliographic format (**IFLA, 2008b**, Appendix C).

In a general analysis, based on experience in the cataloguing treatment of cartographic resources, we have concluded that some codes suffer from being too general and others too specific (Table I). This is the case of "cartographer" (code 180), without discrimination of the scope in *IFLA Universal Bibliographic Control and International MARC Core Programme* (**IFLA, 1994a**, Appendix C), which in practice has been used to describe very different roles (drawer, cartographer, compiler or geographer, surveyor, and even map editor) in documents of different periods of the history of cartography. In contrast, multiple codes can be assigned to the engraver — engraver, etcher, etc... (350, 360, 510, 530 and 760) —, distinguishing, after all, processes that, in order to be properly identified, require technical expertise that is difficult to guarantee in most of the libraries or map libraries.

Table I. Some relator codes in UNIMARC: bibliographic format (based on the Portuguese edition, 2008) and its use recommended or questionable in cartographic resources.

Code	Function	Application	Use in cartographic resources
000	Undetermined function	A person or organization with responsibility undetermined. Used when it is necessary to clarify that an attempt was made, unsuccessfully, to determine the type of responsibility.	While better avoided, this code may be used.
020	Annotator	Writer of manuscript annotations in a printed book.	Responsible for handwritten notes to the document, be it handwritten or printed.
050	Assignee	Person or organization to which a license for printing or publishing has been transferred.	Person or institution responsible for editing.
070	Author Co-author	Person or corporate body chiefly responsible for the creation of the intellectual or artistic content of a work. When more than one person or body jointly bears such responsibility, this code may be used in association with as many headings as is appropriate.	Commonly used for authors with different functions, ranging from those who coordinate, compile, edit, or even draw the map. It is recommended to be replaced by specific relator codes, which best describe the responsibility of each of the production processes.
100	Bibliographic antecedent	One who is the author of the work upon which the work reflected in the catalogue record is based in whole or in part. This relator may be appropriate in records for adaptations, indexes, continuations and sequels by different authors, concordances, etc.	Sometimes used for the author of the map on which the present document is based, but not always to be recorded in 7--.
130	Book designer	Person or corporate body responsible for the entire graphic design of a book, including arrangement of type and illustration, choice of materials, and process to be used.	Sometimes improperly used for the designer. We recommend its use for the coordinator of the technical work of vectorization and digital processing.
180	Cartographer	(Without explicit application)	Function vague, difficult to apply in many cases, designating one who executed the map. Generally attributed both to those who did the surveys or compilations, such as who directs, or draws sketches.

Code	Function	Application	Use in cartographic resources
205	Collaborator	Use only when a more precise function, represented by another code, cannot be used.	Should be used when referring to someone who assists the work of others, exercising a function similar to this. Sometimes improperly used for who raises the map (such as "employee" of the director or coordinator, who directs).
206	Responsible for the collection of field material	Used to person who collect, in the field, elements of musical tradition, oral or naturally related to a particular territory, population or species, transcription, or sound recording or audiovisual.	Should be used for those who perform topographic work or who, somehow, collect information in the field, whatever its nature. It is not generally employed, it is preferred to use the function
220	Compiler	One who produces a collection by selecting and putting together matter from works of various persons or corporate body. Also, one who selects and puts together in one publication matter from the works of one person or body.	One who produces a cartographic document from diverse information, cartographic or not, who brings together and systematizes. It is not generally employed, it is preferred to use 180.
300	Director	Person responsible for the general management of the work or who supervised the production of the performance for stage, screen, or sound recording.	Person who by reason of his functions is responsible for the cartographic document or the one who coordinates or directs the staff.
340	Editor (USED BY: coordinator)	One who prepares for publication a work not his own. The editorial work may be either technical or intellectual.	**Cartographic Editor:** the one who prepares, technically or intellectually, who edits a document that is not of their/ his own. Often used improperly, either to the coordinator or to who runs the surveys.
350	Engraver (USED BY: engraving needle, engraver music and text)	(Without explicit application)	To be used when the statement is generic and when there is no information to select 360, 510, 530, or 760.

Code	Function	Application	Use in cartographic resources
360	Etcher	See Code 350	
400	Funder (obsolete)	USE SPONSOR (723)	Sometimes used, although it is debatable.
410	Graphic technician	USED BY: colorist, graphic designer, designer, printer of photographs and computerized technical drawing. Used for the person responsible for conducting original graphic design (not including those who conceives, the illustrator, code 440).	Includes cartographer-draftsman (or cartographic draftsman) and computerized technical drafting (although these should be separate functions and descriptions that best reflect the activities).
475	Corporate body editor	Corporate body under whose auspices the bibliographic resource may be published. It may be or not the intellectually community responsible for the resource, and whether or not the publisher of the same trade.	Could fit only to institutions that edit the individual works.
490	Licensee	Original recipient of right to print or publish.	Could be applied to cartographic publishers that, until the nineteenth century, obtained from the king the privilege of printing. For modern cartography this could be applied to the commercial publisher, or institution that holds the license for copyright.
510	Lithographer	Person who prepares the stone or grained plate for lithographic printing, including a graphic artist creating an original design while working directly on the surface from which printing will be done.	One who creates the original design on the lithographic stone, from which the reproduction was made. Frequently used in conjunction with 360
530	Metal-engraver	See Code 350	
610	Printer	Printer of texts, whether from type or plates (e.g. stereotype).	Whoever reproduces, usually on paper, the document, regardless of the process that led to the original matrix. The printer is not always recovered in 7 -. Also used for the engraver/printer.

Code	Function	Application	Use in cartographic resources
640	Proof-reader	Corrector of printed matter only; for manuscripts use Corrector (270).	Can adapt to whoever conducts a graphics review and quality control.
700	Copier – USe Scribe	Person who copies manually printed documents, and also an amanuensis or a writer of manuscripts proper	Could be used for cartographer-draftsman, where it is known that he just copied an existing document.
720	Signer	Use for signature which appears in a book without a presentation or other statement indicative of provenance.	Could be used when there is a handwritten signature on the printed document or, if the document is handwritten, it does not correspond to the author or drawer.
760	Wood-engraver	See Code 350	
770	Writer of accompanying material	Writer of significant material which accompanies a sound recording or other audiovisual material.	Should apply to authors of explanations, memoirs, information or notes, together with other codes whenever there was an intervention in the cartographic document (and the record is single).

In the case of cartographic resources, in contrast, the authors and co-authors should be followed by identifying specific functions, omitting the vague code 070. In some cases it happens that, in a document that distinguishes various contributors with different roles, none of them is considered to be the author. Assigning functions to the access points should allow credit to be given to those who perform these functions, whether as the primary or secondary author, and should not exacerbate the confusing issues related to serialization of statements of responsibility in bibliographic records and in the corresponding authority records.

To show that the functions discriminated in UNIMARC should be modified, such that these types of resources can be applied unambiguously, the most frequently used ones have been selected and their application discussed in Table I. The codes chosen to appear in the table are a reflection of the practice of the cataloguing of cartographic resources produced in different periods. On the one hand this shows the inadequacy of some codes and procedures involved in the production of maps until the advent of digital mapping, on the other hand it advances the possibility of adapting some of these codes to the current production, based on new technologies.

In regard to cartographic resources earlier than the nineteenth century, a good example of the contradictory information that can be generated through the attribution of the degree of responsibility and relator codes to access points of a document is the way in which William Faden (1749–1836) appears to be identified: in the databases there are different functions assigned to him, and also their level of intellectual responsibility is different, even when the documents contain statements like "published by W. Faden" or "engraved and published by William Faden" (Table II). Since he was primarily a trader of maps, the allocation to him of primary responsibility in some documents or the choice of relator codes (360, 530 and 650, the selected examples) is disputable.

Table II. Degree of responsibility and functions attributed to William Faden in some records from the National Library of Portugal (BNP) and the National Library of Spain (BNE)

Abbreviated bibliographic description	Intellectual Responsibility Block and attributed relator codes BNP	Intellectual Responsibility Block without relator codes BNE	Information about the responsibility placed on the map
Spain and Portugal divided into their respective kingdoms and provinces, from the Spanish and Portuguese Provincial maps/published by W. Faden, Geographer to His Majesty and to H. R. H. the Prince of Wales. – London : W. Faden, [1796].	700 $aFaden,$bWilliam, $f1749-1836	700 $aFaden,$bWilliam, $f1749-1836	Published by W. Faden, Geographer to His Majesty and to H. R. H. the Prince of Wales Pub.d by W. Faden, Feb. 17, 1795
A new chart of the North coast of Brazil from Seara to the Island of Sn. João Baptista/surveyed by order of the Portuguese government ; Joze Patriceo, Pilot. – London : W. Faden, Geographer to His Majesty, and to His Royal Highness the Prince of Wales, 1809.	700 $aPatrício,$bJosé,$ffl. 1809 702 $aFaden,$bWilliam,$f1749-1836$4650		Surveyed by order of the Portuguese Government Joze Patriceo, Pilot London: Published by W. Faden, Geographer to His Majesty and to His Royal Highness the Prince of Wales. Charing Cross June 4.th 1809.
A topographical map of the isle of Minorca... / by L. S. de la Rochette ; grav. William Faden. – London : William Faden, 1780.	700 $aDelarochete,$bLouis Stanislas d'Arcy,$f1731-1802 702 $aFaden,$bWilliam,$f1749-1836$4360	700 $aDelarochete,$bLouis Stanislas d'Arcy, $f1731-1802 700 $aFaden,$bWilliam, $f1749-1836	Geometrically Surveyed by the Royal Engineers, while it remained in the possession of the French during the Last War, and digested by L. S. de la Rochette Engraved and published by William Faden, Charing Cross as the Act directs Jan.y 1.st 1780

Abbreviated bibliographic description	Intellectual Responsibility Block and attributed relator codes BNP	Intellectual Responsibility Block without relator codes BNE	Information about the responsibility placed on the map
A chart of the coasts of Spain and Portugal with the Balearie Islands and part of the coast of Barbary/ by L. S. De La Rochete ; engraved and published as the Act directs by William Faden. – London : William Faden, August 30th 1779.	700 $aDelarochete,$bLouis Stanislas d'Arcy,$f1731-1802 702 $aFaden,$bWilliam, $f1749-1836$4530	700 $aDelarochete,$bLouis Stanislas d'Arcy,$f1731-1802 700 $aFaden,$bWilliam, $f1749-1836	by L. S. de la Rochette Engraved & published as the Act directs, by Wm Faden, success.^{r} to the late T. Jefferys, Geog.^{r} to the King. Charing Cross, August 30.th 1779.
Chart of the entrance of the river Tagus/surveyed in 1806 by William Chapman, Master of the Royal Navy ; writing by T. Harmer ; published by W[illia]m Faden, Geographer to His Majesty and to His Royal Highness the Prince of Wales ; H. Wilson sculpt. – London : W[illia]m Faden,12th August, 1807.	700 $aChapman,$bWilliam, $f1749-1832 702 $aFaden,$bWilliam, $f1749-1836$4650 702 $aHarmer,$bThomas, $ffl. 1784-1814$4530 702 $aHarmer,$bThomas, $ffl. 1784-1814$4530 702 $aWilson,$bH., $ffl. ca 1807$4530		Surveyed in 1806 by William Chapman, Master of the Royal Navy. Published by Wm Faden, Geographer to his Majesty and to his Royal Highness the Prince of Wales, Charing Cross, August 12.th 1807. Writing by T. Harmar H. Wilson sculpsit.

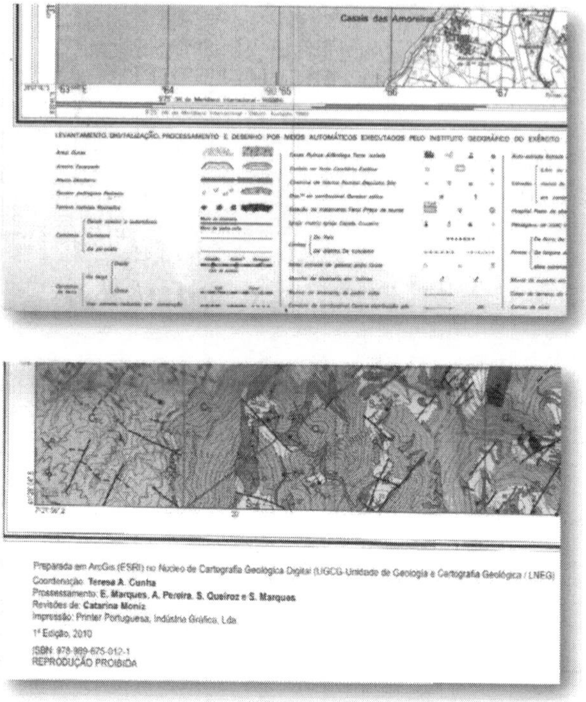

Figure. 3 Examples of indication of the statement of responsibility and associated functions in recent maps: Carta militar de Portugal 1:25,000 from Instituto Geográfico do Exército (above) and Carta geológica de Portugal 1:50,000 from Laboratório Nacional de Geologia (below).

Apart from the generic codes listed at the beginning of this analysis, contemporary cartography does not raise insurmountable problems in using the existing codes. After applying some adaptations, it should be possible to use the codes defined in the *UNIMARC manual*. The analysis of some topographic and thematic maps produced in Portugal with the use of digital tools (Figure 3) shows that new functions — inherent to electronic processes — are being added to the traditional ones. Examples are: "scanning, processing and design by automated means" or the indication of specific tools of Geographic Information Systems (GIS), associated with terms of responsibility for the coordination, processing and review procedures that were used. Thus, contrary to the old maps, the application of relator codes does not seem to raise issues that are sufficiently relevant to require the need for a complete restructuring of the codes.

The potential of the application of these codes for cataloguing systems management (documentation) and for its users (added value for the retrieval and use of information) is important to consider when examining the related technical issues. An enquiry was made among non-specialist users of Portuguese cataloguing systems about the potential of properly applied relator codes for the efficient management and retrieval of information. The responses show that they consider them to be useful for producing bibliographical lists of a given author (a physical person or an organization), with a differentiation by his intellectual responsibility in the work.

The analysis of this aspect in some national and international bibliographic databases, with regard to cartographic resources, also demonstrates some problems in filling out this subfield, as exemplified in the records extracted from the British and French national libraries (Table III). In these records the problem of identifying and selecting the principal author, the co-authors and the secondary statements of responsibility is well illustrated, because all contributions were recorded in the same field 7, even though each of them has played a different role. Also, different relator codes (650 and 070) were attributed to the same author, in spite of the fact that his contribution (editor) for the two maps had been identical.

The example of the British Library, where the relator codes do not seem to be used, allows us to confirm the low value associated with this information and, therefore, to question the continuation of its mandatory use in most Portuguese cataloguing systems despite the optional nature allowed by the format.

For users of cartographic documents, the importance of the information about the nature of the contribution of a particular author to the contents of a map depends on the knowledge that these users possess. For those mainly interested in the geographic content, some maps may be used to study the evolution of cartography as a science, and their authors may be related to periods and events that marked the history of a territory. Such information

Table III. Records extracted from the National Library of France (BNF) and the British Library (BL).

NATIONAL LIBRARY OF FRANCE
2001 $aA plan of the city and environs of Philadelpia$bDocument cartographique$fsurvey'd by N. Scull and G. Heap$gengraved by Will. Faden 210 $a[London]$cFaden$d[1777] 300 $aCarte numérisée à la Library of Congress 300 $ahttp://hdl.loc.gov/loc.gmd/g3300m.gar00002 607 315238447aPhiladelphie (Pennsylvanie, États- Unis)$2AFNOR NF Z 44-081 608 $aCarte$2frTCP 660 $an-us-pa 702 |$315150659$aScull$bNicholas$f17..-17..? $ccartographe$4180 702 |315138256aHeap$bGeorge$f17..-17..? $cgéographe$4180 702 |315615346aFaden$bWilliam$f1749-1836$4350 702 |$315615346$aFaden$bWilliam$f1749-1836$4650
2001 $aA Plan of the town of Newport in Rhode Island$bDocument cartographique$fsurveyed by Charles Blaskowitz$gpublished by W. Faden, 1777 210 $a[S.l.]$c[s.n.]$d[1887] 300 $aT. VI. 2, p. 597 de Winsor 607 315239401aNewport (Rhode Island, États-Unis) 315231655z1777$2AFNOR NF Z 44-081 608 $aCarte$2frTCP 660 $an-us-ri 702 |$315129949$aBlaskowitz$bCharles$f1760-1823 $4070 702 |$315615346$aFaden$bWilliam$f1749-1836$4070 702 |312192902aWinsor$bJustin$f1831-1897$4070

BRITISH LIBRARY	
24512	|a A map of the British Colonies in North America, with the roads, distances, limits and extent of the settlements ... inscribed to ... the Earl of Halifax [with] The British Colonies in North America, engraved by William Faden, 1777 / |c John Mitchell.
250	|a [4th ed]
255	|a Scale [ca. 1:2 000 000]. English miles 200 [= 165mm].
260	|a London : |b Publish'd by the author Feb.ry 13.th 1755 according to Act of Parliament : |b Printed for Jefferys & Faden Geographers to the King at the corner of St. Martin's Lane Charing Cross London, |c [17--?]
300	|a 1 map on 8 sheets : |b some hand col ; |c each sheet, 49 x 64cm.
7001	|a Kitchin, Thomas.
7001	|a Jefferys, Thomas.
7001	|a Faden, William.

would therefore certainly be useful for them, but the information given to them not only not acknowledges the institutions, but also contributes to notable errors in this area.

This being the case, it may not be worthwhile using this information, taking into account the difficulty of the assignment based on the table below and the insignificant purposes and results that apparently are expected to be obtained, unless a restructuring is carried out in accordance with cartographic resources that seem to have been forgotten in the preparation of the list integrated in the *UNIMARC Manual*.

4. CONCLUSION

This analysis concludes that a proper treatment of the authors associated with cartographic resources, produced at different times and using different processes, describing contributors with different roles and levels of responsibility, is one of the key pillars to ensure a more effective management of the bibliographic databases and standards of quality of information provided to the users.

The brief description given of the evolution of the cartographic production process shows that treatment of the statement of responsibility in the bibliographic description of the maps requires that cataloguers, in addition to extensive experience in the application of technical standards, have appropriate training enabling them to identify and distinguish these references.

The analysis of the role of relator codes, stipulated by the UNIMARC Manual, leads us to conclude that it is necessary to adapt these codes.

Until this happens, we suggest that the present codes, assigned in the *UNIMARC manual*on a voluntary basis, should not be used, because we believe that their implementation in the present form only contributes to increasing the number of errors that normally proliferate in the databases.

NOTES

[1]This paper incorporates part of the research carried out for the Master Degree in Library and Information Sciences (see: **Domingues, 2010**).

[2]"Changed from handmade production (or analog) to production based on computer means (or digital) and, consequently, from paper maps to digital maps", as well as, "production, before, almost exclusively centered in professionals (cartographers) and institutions concerning cartographic activities and nowadays more accessible to anyone with minimum technological means".

[3]"The geographer works at home, having at hand several documents of the same area, which he compares and adapts according to what in his opinion is the most perfect reproduction. His job is not to produce specific maps, because

for that there is another "class of people", who do not need any special training apart from knowing rectilinear geometry. If geographers would need to see and measure the area shown by their maps, none would have been able to publish in his lifetime one of the four parts of the earth."

[4]"So, like many other geographers, cartographers and European publishers, Lopez prepares his maps by combination, confrontation and synthesis of the chorographical and geographical descriptions and maps of different scales and types, the so-called office Cartography. The 'other class of people' are the engineers, particularly the military, who represent the 'modern' ones, and who defend maps based on the establishment of triangulation and surveying".

REFERENCES

1. Andrew, P. G., & Larsgaard, M. L. (Eds.) (1999). Maps and related cartographic materials: cataloguing, classification, and bibliographic control. New York: The Haworth Information Press

2. Dias, M. H. (2007). Cartografia temática: programa. Lisboa: Centro de Estudos Geográficos (Geo-Ecologia, relatório n.º 6)

3. Domingues, S. (2010). Os recursos cartográficos e as bases de autoridade: normas e aplicações. Lisboa: [s.n.]. Tese de mestrado, Ciências da Documentação e Informação, Universidade de Lisboa, Faculdade de Letras

4. Garcia, J. C., & Moreira, L. M. (2008). El geógrafo trabaja en su casa: espaços portugueses na produção cartográfica de Tomás López. Península. Revista de Estudos Ibéricos, 5, 103–125. Retrieved December 5, 2012 from http://ler.letras.up.pt/uploads/ficheiros/4709.pdf

5. IFLA (1994a). UNIMARC manual: Bibliographic format. Retrieved December 5, 2012 from http://archive.ifla.org/VI/3/p1996-1/appx-c.htm

6. IFLA (2008b). UNIMARC manual: Bibliographic format. Ed. by Hopkinson, A. 3d ed. München: Saur

7. Pastoureau, M. (1980). Les livres et les murs. In Cartes et figures de la terre (pp. 436–454). Paris: Centre Georges Pompidou, Centre de création industrielle

8. Pedley, M. S. (2005). The commerce of cartography: making and marketing maps in eighteenth-century France and England. Chicago: The University of Chicago Press

9. Rockwell, K. (1999). Problem areas in the descriptive cataloging of sheet maps. In P. G. Andrew, & M. L. Larsgaard (Eds.), Maps and related cartographic materials: cataloguing, classification, and bibliographic control (pp. 39–63). New York: The Haworth Information Press. Also: Cataloging & Classification Quarterly, 27, 39–63.

NOTES

[1]This paper incorporates part of the research carried out for the Master Degree in Library and Information Sciences (see: **Domingues, 2010**).

[2]"Changed from handmade production (or analog) to production based on computer means (or digital) and, consequently, from paper maps to digital maps", as well as, "production, before, almost exclusively centered in professionals (cartographers) and institutions concerning cartographic activities and nowadays more accessible to anyone with minimum technological means".

[3]"The geographer works at home, having at hand several documents of the same area, which he compares and adapts according to what in his opinion is the most perfect reproduction. His job is not to produce specific maps, because for that there is another "class of people", who do not need any special training apart from knowing rectilinear geometry. If geographers would need to see and measure the area shown by their maps, none would have been able to publish in his lifetime one of the four parts of the earth."

[4]"So, like many other geographers, cartographers and European publishers, Lopez prepares his maps by combination, confrontation and synthesis of the

chorographical and geographical descriptions and maps of different scales and types, the so-called office Cartography. The 'other class of people' are the engineers, particularly the military, who represent the 'modern' ones, and who defend maps based on the establishment of triangulation and surveying".

Chapter 9

One-To-Many: Building a single-search interface for disparate resources

Cole Hudson, Graham Hukill

Wayne State University,

Libraries provide access to a complex array of resources, but it comes as little surprise that many struggle with this task. The ubiquity of Google in modern research has placed libraries in the position of trying to emulate or set themselves apart from the search engine, regardless of how appropriate it is to compare libraries to Google. There are numerous ways in which libraries can improve the search and discovery experience for patrons, and for many, this improvement currently comes in the form of custom-built or commercial discovery systems—which aggregate disparate library content into a single results display. But, regardless of discovery systems' potential, as Lown, Sierra, and Boyer (2013) conclude, libraries should learn and balance user expectations with the actual capabilities of library information systems. In an effort to find this balance, our work led us to a singular goal: providing a single-search interface for our complex array of library resources. What resulted was a discovery tool we named QuickSearch.

Developed in-house, QuickSearch is a single point of interaction, "bento style" search portal, displaying search results from disparate resources in discrete boxes. These resources come from numerous independent, back-end systems, including search results from the library catalog, Serials Solutions Summon Service, research guides, journals, our institutional repository, digital collections, databases, and a site search of the entire university website. The term bento comes from Japanese cuisine where different parts of a meal are compartmentalized in aesthetically pleasing ways; this ultimately is the aesthetic goal of compartmentalizing search results from disparate resources on the results page.

In this chapter we are not advocating for specific programming languages or technologies; instead, we aim to share our underlying design principles and philosophy in developing QuickSearch. As we see discovery moving toward constantly shifting silos of data that must be aggregated for users, we tried to imagine an architecture that accommodates this reality.

Vendors (e.g., Serial Solutions, EBSCO) offer products that aggregate library resources, but as we shall see, these products are, by nature, incomplete and incomprehensive. We are unable to rely on vendors for a single point of search, yet we are also unable to create our own single resource index. The discovery landscape is changing too quickly, and our staff is too small for a monolithic approach. Instead, we have chosen to address data silos individually and loosely couple them together under a single banner: QuickSearch—software that our small team of developers can update and maintain. This approach seems to be well received by users and has already resulted in increased access to resources.

This chapter will first outline QuickSearch's key features (see figure 11.1), discuss explored and realized approaches, and conclude with an appraisal of the impact QuickSearch has had on user behavior patterns.

Potential Approaches And Why They Didn't Work

To meet our goal of providing a comprehensive, intuitive search tool across all library resources, from a single interface, we first explored a range of different possible approaches. This section will discuss these, noting why they were not a good fit for us, in an effort to contextualize the development of QuickSearch and explain how it addressed problems with these other approaches. (See table 11.1.)

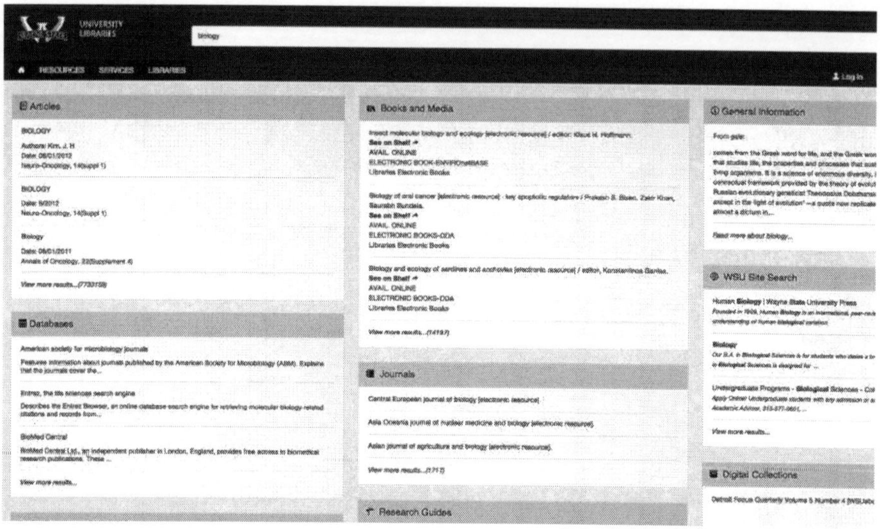

Figure 1. Screenshot of QuickSearch displaying results for "biology" search

Table. 1 Overview of explored approaches

Approach	Problem	QuickSearch's Solution
Web-Scale Discovery Services	Do not contain all library resources	WSDS results are valuable, incorporated as "Articles" box
Single Index of Library Resources	Difficult to rank, difficult to maintain	Currently avoids ranking
Monolithic Software Approach	Brittle, hard to update, too much maintenance	Simply designed, low barrier to construction, easy to update, resilient to catastrophic failure

Web-Scale Discovery Services

Providing a single point of interaction for search is increasingly a focus of vendors, as they realize how valuable this would be to libraries and their users. Companies such as EBSCO, Ex Libris, OCLC, and ProQuest all have products that index multiple resource types, including databases that they operate and provide access to, abstracts from other vendors' databases, and even MARC records for books and media from a library's catalog. These platforms are often referred to as "web-scale discovery services" (WSDS). The number of distinct items in these platforms is staggering—our own ProQuest "Summon" platform returns 632,286,219 results for a blanket search. Yet still, some library resources are not included, making them effectively invisible to our users.

Web-scale discovery services include resources from a variety of sources. They are well suited for searching individual articles because they include articles from a wide variety of journals and databases, in addition to MARC records they pull from library catalogs and other resources. But in spite of the increasingly wide net that web-scale discovery services cast over a library's resources, vast swaths are often missed: institutional repositories, digital collections, databases, research guides, information on the university or library website, and so on. These omissions can happen for a variety of reasons, three of which we will explore further below.

One reason resources might be absent from a WSDS is that they don't fit neatly into the model the service is using for indexing and searching resources. As advanced and sophisticated as they get, web-scale discovery services still rely on quasi-traditional approaches to modeling library resources: titles, authors, dates, publishers, and so forth. For example, they are ill prepared to accommodate a university's website search results where complex, algorithmic ranking of pages is useful, if not essential. Library research guides are another example of resources that do not fit neatly into that model. Web-scale discovery services are successful only insofar as the resources we seek can be described and searched in a meaningful way through their interface.

Second, it is possible the resources are inaccessible to the WSDS, or too small and unique to be indexed by a large, vendor-created WSDS. For example, at Wayne State University we have a growing institutional repository (DigitalCommons@WayneState) and online digital collections, neither of which are indexed by our Summon platform. These are valuable library resources, but are missed completely.

Thirdly, even within resource types that a WSDS is able to index and search, there might be gaps when resources are not included due to "reluctant vendor participation/partnerships or to choices of resource inclusion made by libraries" (Ellero 2013). In effect, the coverage of a WSDS is a moving target, which does not always support comprehensive, repeatable searching.

The point here is not to provide an in-depth analysis of web-scale discovery services, which other articles have expertly done (e.g., Ellero 2013), but instead to show that despite their vast reach, there are library resources they currently cannot, and perhaps will not, index for search.

Single Index: Apples And Oranges

Another approach to providing a single-search interface for library patrons is to put metadata records from disparate sources into a single index, and provide a search interface from that single point. Cosmetically and conceptually it is quite similar to Google, in that all resources would come from a central index, interleaved into a single list of results. This approach looks elegant on paper, but is fraught with its own problems. A particularly thorny one is that

library data just doesn't have the same characteristics as Web data. It just doesn't. Though . . . lots of folks working on Web Scale Discovery have made big strides, nobody has "solved" relevance ranking for full library discovery the way Google solved it for the Web. (Thomale 2015)

This quote from Jason Thomale during his talk at Code4Lib Annual in 2015 (also see chapter 10, "The Bento Box Design Pattern," by Thomale,

Philipps, and Hicks) very nicely sums up the problem of creating a single index, comprised of multiple resource types, to search and provide results: the resource types that are included are simply too different, and ranking them as search results is an effort in futility. Search algorithms that power a single database or resource type are complex at best, but ranking multiple resource types increases that complexity exponentially.

With the goal of a single-search box interface, a bento-style approach sidesteps some problems with this approach, namely it does not purport to rank variant resource types. Instead, it provides these resources on equal footing, leveraging the ranking algorithms internal to each resource's individual search interface.

Monolithic Software Application: Less Code, More Uniformity

A third approach is to design a system in which resources share common pathways and code. Each "box" on the page has to ask its associated database for data in a particular fashion. Each database returns this data in a particular form. Given five to ten boxes, the complexity of code needed to handle these differences can grow quite dramatically. The way to prevent this would be to pass everything through a common processing pipe that would handle requesting and returning data to the "boxes" on the page. This approach would allow for less overall code and more uniform features between the boxes. If every box uses the same processing pipe, potentially when an update to one box is applied, it would apply to all. This approach, however, is not without its pitfalls.

The main issue with this kind of approach can be best expressed in its need for dedicated staff to manage the system. In a 2001 IBM white paper, Paul Horn states that the greatest obstacle facing the IT industry is complexity. "In fact, the growing complexity of the I/T infrastructure threatens to undermine the very benefits information technology aims to provide." This problem is one of scale, because soon, he warns, there would not be enough IT professionals to handle the rise in complex, interconnected systems (Horn 2001). Dystopic vision aside, Horn does touch upon a key issue that applies to any organization regardless of size—the increased human cost of ever more complex systems. Horn's premonitions are relevant to our work with QuickSearch. Maintaining a more complex and better-integrated system would be cost-prohibitive in terms of dedicated staff time. We have no full-time front or back-end web developers.

The main development team consists of librarians and system administrators, for which application development is only a part of their job duties. If we followed a monolithic development approach, one in which we wrote DRY (Don't Repeat Yourself) code, this would naturally lead to something such as a common processing pipe for all resource types, but the costs of maintaining this approach would increase.

Another issue to contend with is that the resources which power QuickSearch undergo changes and upgrades at different rates either because of (1) internal decisions to change how we interact with a resource, or (2) changes to resources outside of our control (such as Serial Solutions' Summon service or Springshare's LibGuides API). When the resources share a common processing pipe, any changes we make to one resource would have to not disrupt the functionality of other resources. This, in turn, increases the amount of testing needed to ensure that all the resources still retrieve and display results correctly. With enough people and very established application development workflows, interconnected systems can be run smoothly. With our small team, this was a major roadblock to adopting this overall approach.

Quicksearch: Our Approach, Our Solution

Given our goal of a single-search interface across all library resources, we developed QuickSearch with the shortcomings of other approaches firmly in mind, focusing on a low-barrier, manageable approach we hoped would hit all of our requirements.

The design principle is straightforward: QuickSearch is a representative specimen of the bento box-style search interface, where results from different sources are returned to visually discrete boxes on the results page. One page, one search box, results from as many library resource types as possible.

Design Principle: Every Box For Itself

As opposed to the monolithic approach discussed above, we opted for an architecture with a much lower barrier to entry. Figure 11.2 is an overview of the QuickSearch architecture.

Our strategy going into development—when the number of resources types (boxes) was still fluctuating—was to have each "box" on the search results page have its own "pipe" back to the original source that provided results relevant to the current search. The advantages to this approach have been numerous.

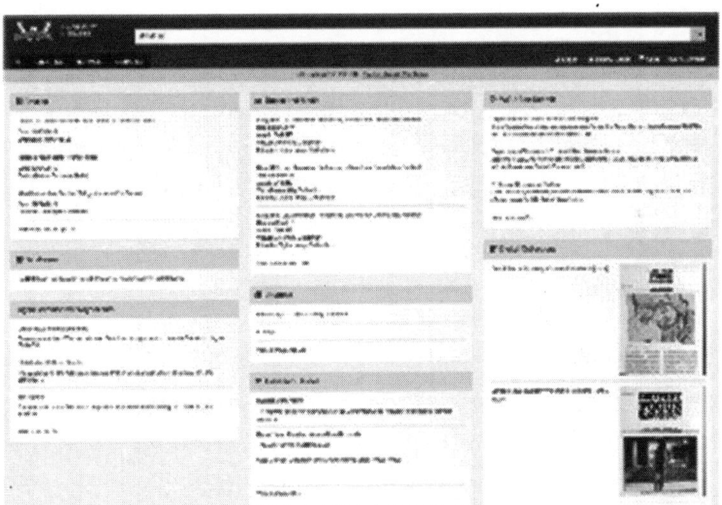

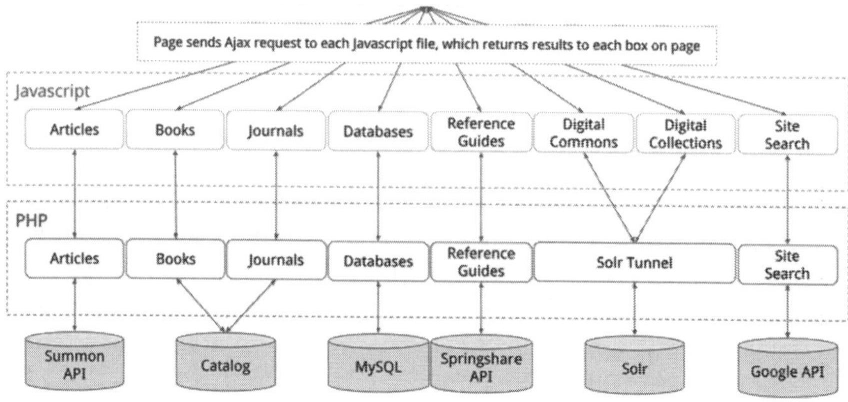

Figure 2 Architecture of QuickSearch.

Low Barrier To Creation

Even during the process of identifying and configuring data sources that would populate the search results, we could start building "pipes" for known resource types and could begin prototyping the page. In a monolithic-based approach, it might have been cumbersome to wait on unknowns like this, such that we could fold them into more terse and purposeful code.

Easy To Comprehend

Just follow the arrows! Most librarians who work with electronic resources will be familiar with the databases at the bottom of the figure (some are vendors, some are databases with indexed resources). Except for a couple of unique cases where different "boxes" share "pipes," there is a near 1:1:1 ratio from JavaScript-populated "box," to PHP "tunnel," to database on the back end.

Well Suited For A Small Team

The distributed nature of the design—highly modular and autonomous routes for each "box"—made it easy to develop with a small number of people working on the front and back ends. In many instances, we split up the work by resource type, where each person would code the "pipe" for a given resource from beginning to end. They would be responsible for the actual database queries, moving up through PHP, JavaScript, and finally into displaying results. In a larger, more monolithic system, this kind of workflow would be nearly impossible. The only constraint we had was that box results must be reliable and consistent with the resource's native results set; the mechanics of each "pipe" were determined by the nature of the resource database.

Nuts And Bolts Overview Of Quicksearch

Aside from the traditional markup and styling in HTML and CSS, the application consists of JavaScript- and PHP-mediated interactions with HTTP-accessible data end points. JavaScript renders resource data onto a user's page, while PHP communicates with the data resource. Each resource has its own JavaScript and PHP scripts which handle querying and response separately from the other available QuickSearch resources.

When a patron first uses the QuickSearch tool, their query term(s) are captured via JavaScript. AJAX (Asynchronous JavaScript and XML) calls, each connected to a separate box on the page, then pass along the data for processing by the server. Server-side scripts written in PHP take care of formatting and sending query terms to their corresponding database. When the database responds back with the patron's requested resources, the PHP script processes and sends the data back up the pipeline. Finally, the data is passed back to the QuickSearch web page where it is inserted into the database's corresponding box. Due to the built-in asynchronicity of the AJAX calls, which handle the patron's input and the database's corresponding output, each set of data is returned only when it is ready and the resource's failure and speed (or lack thereof) has no impact upon the other resource boxes or their ability to load data from their own data sources.

Impact And Engagement

QuickSearch was launched November 6, 2013, as an additional search option among a cluster of search interfaces. Though it provided an umbrella search that covered resources from other search interfaces, it was not until August 2014, when we launched a redesign of the website, that we did away with the search interface cluster approach and featured QuickSearch as the primary search interface from the library home page.

Since that time, inroads for search and discovery have changed dramatically. What used to be unique and isolated search interfaces have become a single point of interaction. The result has been largely positive. Users have demonstrated active and sustained engagement on the search page; resources that were previously unsearchable and inaccessible from the library website are now discoverable; and we received a healthy dose of "No News is Good News" feedback from users.

INCREASED ACCESS TO RESOURCES

A driving force for the creation of QuickSearch was to provide discoverability to resources that were previously unsearchable from the library website. Nevertheless, many of the resources now searched by QuickSearch were, in fact, previously searchable from the library website via a tabbed search interface on the library home page: Articles ("Everything"), Books and Media ("Catalog"), Databases ("Article Databases"), Journals ("Online Journals"), and even Site Search.

Though all of these resource types were searchable, tab usage was inconsistent, and search was certainly not possible across all resource types at once. The effect hurt serendipitous discovery across resource types, and required a duplication of effort on the user's part, constantly repeating searches in different interfaces, navigating the challenge and peculiarities of each.

With the launch of QuickSearch, a new tab was created allowing users to search across these resources with one search (see figure 11.3).

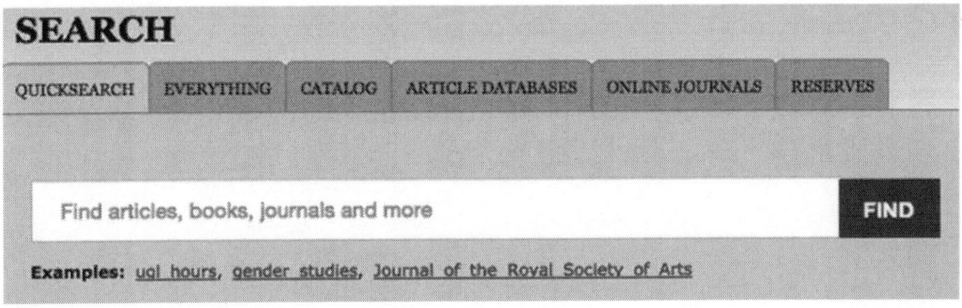

Figure 3 Pre-redesign website tabbed search interfaces with QuickSearch included.

Figure 4. Current website without tabbed search interfaces, featuring QuickSearch as the sole search box in header.

Even with the addition of QuickSearch, tab usage was still inconsistent and confusing, and patrons still had the option to search specific resource types via their isolated search interface. With the redesign in August 2014, we moved away from the tabbed search cluster, and for the first time, started pushing all library search and discovery traffic from our new standard header (see figure 11.4) through QuickSearch.

As user searching began routing through QuickSearch only, many library resources that were previously hidden from search and discovery on the main library website were, for the first time, exposed to all searches, for all users. This includes our Institutional Repository (DC@WSU), our Digital Collections (WSUDOR), Library Research Guides, and a more sophisticated Site Search of the entire wayne.edu domain.

This exposure has resulted in increased use and awareness of these resources. Since QuickSearch launched in August 2014, through March 2015, it has directed users to resources previously unsearchable from the library website:

- 800 + visits to digital collections

- ` 3,800 + visits to library research guides
- ` 3,500 + visits to DC@WSU

Those visits to DC@WSU, as an example, are particularly encouraging. DC@ WSU is a Software-as-a-Service (SaaS) from Bepress, with a website and search interface entirely removed from the library infrastructure. Usage of this resource was limited to the traffic we could direct there, often just for known items such as electronic theses or dissertations. These near 4,000 visits not only demonstrate increased usage of DC@WSU's valuable, Wayne State scholarship, but have produced increased awareness around DC@WSU and its mission as the university's institutional repository. Increased campus awareness of DC@WSU is a major, somewhat unforeseen benefit of simply showing up on the QuickSearch results page. The same can be said about Digital Collections, Research Guides, and other parts of the university website that Site Search scours.

USER ENGAGEMENT BY THE NUMBERS

Just shy of 300,000 visits (where a "visit" is defined as a new user visiting the page, or a returning user after thirty minutes or more) at the time of this writing, QuickSearch represents a substantial portion of all traffic to the library website, which had approximately 1,600,000 visits in this same time period. But even more interesting than simple visits to the page is user behavior once there.

To collect statistics on QuickSearch, we use an open-source website analytics platform, Piwik (http://piwik.org/). We tracked visits to the page, searches on the page, and clicks to actual resources. While "outlinks" (in Piwik nomenclature) may contain clicks to the header menu items, we believe the vast majority of those clicks are to QuickSearch search results. We have recently started quantifying only QuickSearch clicks; this will allow us greater granularity in future analysis. From this tracking, a couple interesting insights emerge.

First, use of QuickSearch was minimal until it was featured as the primary library search interface; even though it was offered as a tab on the search interface cluster, and at one point even the default one, it was not widely used. Traffic increased sharply around the beginning of the fall semester after the redesign. For our team, this slow rollout was a good time to identify and fix bugs, while traffic was slow. Having other librarians test the system by performing searches and noting irregularities was instrumental during that time.

Second, and perhaps the most interesting insight to emerge from the numbers, is the behavior of users on the page. Reviewing the analytics have identified a rough 1:2:4 ratio of visits: searches: resource use.

Piwik's web logs serve as a cross-check. We can look at anonymized individual user visits to see the activity in search terms and links out. While these do not perfectly model the 1–2-4 ratio we see in the graphs, they serve as sketches of common user behavior that leads to the patterns we observe among users overall.

QuickSearch demonstrates a high level of user engagement. By contrast, even a website such as our LibGuides platform—with plenty of dynamic content—has a much closer 1:1:1 ratio of visits, activity, and outlinks.

We believe we can assert that users are engaged with the website, performing and refining queries, and perhaps most importantly, are being exposed to a large and diverse array of library resources. While we cannot make inferences about the relevance of search results for our users at this time, we are happy with this level of engagement. Understanding user engagement with QuickSearch in a more comprehensive way is a next step as we continue to develop and refine this tool.

"NO NEWS IS GOOD NEWS"

In thinking about the impact QuickSearch has had on discovery for our patrons, we spoke with our user experience librarian who has been operating a survey since the launch of QuickSearch and the website redesign. The survey was provided to users as a banner across the top of QuickSearch. While we initially lamented only 45–50 survey responses regarding QuickSearch, this librarian astutely pointed out that considering more than 300,000 QuickSearch searches, only 45–50 responses, and the somewhat unfortunate reality that most feedback is negative when it does come back, one might chalk up a small response rate to "No news is good news," and assume it quietly became a successful and functional search interface for users. If things were not working well, we most likely would have heard about it.

Further user data supports this as well. Figure 11.5 represents returning visitors to QuickSearch. Over the course of a semester, the number of returning visitors increases. While QuickSearch is our primary search interface now, other more targeted search interfaces do still exist, even featured prominently as "QuickLinks" on the front page, yet users continue to return to QuickSearch to perform their searches.

Given these markers, and some anecdotal testing and feedback from users, we feel confident that QuickSearch has integrated itself as the primary search interface for our users, with a high degree of use and engagement, providing discoverability and access to a wider array of library resources than ever before, from one location.

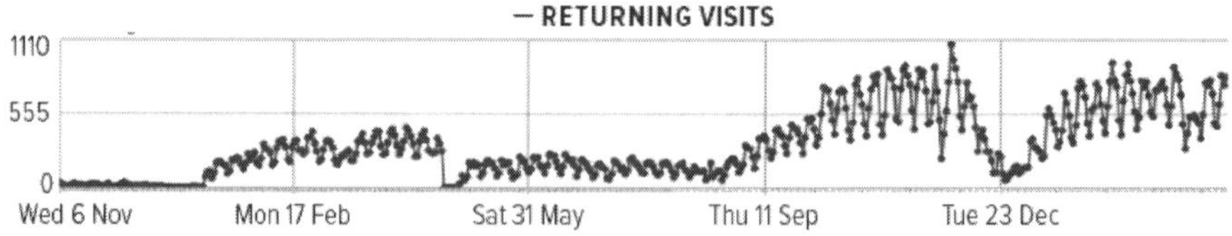

Figure4. Returning users to QuickSearch (visits triggered by user searches)

CONCLUSION

In building QuickSearch, we are confident that we met our goal of providing a service that unites our disparate resources. It is sustainable for our development team, and we see the continued and sustained use by our patrons as positive. However, regardless of the tool's current success, we realize that user expectations about resources and their presentation will evolve. Even now, we are evaluating radical changes to QuickSearch's interface. Regardless of how it evolves, we are confident the design principles that underlie QuickSearch—modular components, ease of development for a small team, and flexibility towards resource types—will remain constant as we move our discovery efforts forward.

ACKNOWLEDGMENTS

Building QuickSearch was a collaborative effort within the library. Development would not have been possible without the feedback and input of our librarians and staff. Particular credit and thanks to Rachael Clark, Joseph Gajda, Niranjan Jadhav, Axa Mei Liauw, Joshua Neds-Fox, Elliot Polak, Vinay Potluri, and Negib Sherif.

REFERENCES

1. Ellero, Nadine P. 2013. "Integration or Disintegration: Where Is Discovery Headed?" Journal of Library Metadata 13, no. 4: 311–29.
2. Horn, Paul. 2001. Autonomic Computing: IBM's Perspective on the State of Information Technology. Armonk, NY: IBM.
3. Lown, Cory, Tito Sierra, and Josh Boyer. 2013. "How Users Search the Library from a Single Search Box." College and Research Libraries 74, no. 3: 227–41

4. Thomale, Jason. 2015. "You Gotta Keep 'em Separated: The Case for Bento Box Discovery Interfaces." Lecture presented at the Code4Lib conference, Portland, Oregon, on February 11, 2015.

Chapter 10

Impact of Knowledge Incentive Mechanisms on Individual Knowledge Creation Behavior—An Empirical Study for Taiwanese R&D Professionals

Min-Shi Liu

Department of Business Administration, Soochow University, Taipei 100, Taiwan

ABSTRACT

Knowledge incentive mechanisms are a new development in the management of knowledge. The most prevalent perspective presented in the available literature is that the knowledge incentive mechanisms must have the ability to manage the flow of business knowledge. This study explores the relationship among knowledge incentive mechanisms, knowledge psychological ownership, and individual knowledge creation behavior. This study uses structural equation modeling to test a sample of R&D professionals from high-tech companies in Taiwan. Data analysis suggests that knowledge incentive mechanisms promote the psychological ownership by the knowledge owner so that he or she actively carries out knowledge innovation. This implies that knowledge incentive mechanisms stress private proprietary characteristics employees who possess real ownership when they innovate will deem such innovation as an extension of self, giving rise to a protective or defensive mindset, which then has an impact on individual knowledge creation behaviors because of the psychological ownership of knowledge.

Keywords: Knowledge incentive mechanisms; Knowledge psychological ownership; Individual knowledge creation behaviors

1. INTRODUCTION

Knowledge-based view (KBV) researchers increasingly have paid attention to incentives in realizing the innovation-related activities, where effort and learning have a high degree of complementarity (Bridoux et al., 2011, Coff, 2003, Insead and Lim, 2007 and Ziedonis, 2004). Their main argument is that knowledge-based (value creation perspective) and incentives-based (value appropriation perspective) perspectives complement each other to explain individual knowledge creation behaviors. In fostering knowledge creation behaviors among employees, managers usually face a public good dilemma: the maximization of individual employees' material payoffs conflicts with the achievement of the collective goal of creating knowledge for the firm (Fehr and Fischbacher, 2002 and Ostrom, 2000). In recent work, scholars have discussed the diversity of incentives driving individual knowledge creation behaviors (Gottschalg and Zollo, 2007 and Nickerson and Zenger, 2008), but these studies have not explored the mechanisms to integrate both perspectives.

Knowledge incentive mechanisms are the newest trend in corporate knowledge management (KM), and recent research has come up with a number of different definitions of such knowledge incentive mechanisms (Fey and Birkinshaw, 2005, Foss, 2006, Grandori, 2001, Hsu, 2006, Nickerson and Zenger, 2004 and Osterloh and Frey, 2000). Child and McGrath (2001) proposed that organizations based on manpower capital are different from traditional enterprises in that they rely less on power hierarchies and high-powered performance incentives, and instead focus on culture and clan controls. But Teece

(2003) put forward the opposite view that minimum supervision and high-powered performance incentives are in fact effective tools for enhancing innovative behavior within an organization. No matter what kind of knowledge incentive mechanism they put at the heart of their research, past papers mostly agree that any such mechanism should have the capacity to promote the circulation of knowledge within a business or organization. The problem at this point is the lack of a good theory, and of a way of systematically testing knowledge incentive mechanisms. This research attempts to explore knowledge incentive mechanism from the perspective of property rights (as a form of ownership), which represents an entirely new approach to KM aimed at stimulating organizational innovation.

Another angle from which to deal with knowledge and its management is the stock view, which sees knowledge as a property. Pierce, Rubenfeld, and Morgan (1991) thinks that ownership has a significant psychological effect on employees' behavior and innovative attitude. The term psychological ownership is used in this context to describe a feeling among an organization's employees that a particular target or object (knowledge) belongs to me or us (Pierce et al., 1991), and the theory is that for any knowledge incentive mechanism to effectively promote innovative behavior, it needs to utilize these feelings of psychological ownership. An important question asked in this approach is: how can knowledge incentive mechanisms help to stimulate and motivate owners of knowledge, and reduce the emotional distance to knowledge and ideas, so that knowledge can be allocated and applied in the most effective manner?

In order to integrate and extend these existing views, this study develops a theoretical frame and undertakes an empirical study to explore the knowledge creation behaviors of R&D professionals in Taiwanese high-tech industries. The structure of this study is presented as following. In Section 2, this study reviews the related literature on knowledge incentive mechanisms from a property rights perspective, knowledge psychological ownership, and individual knowledge creation behavior, and present an integrated model followed by the hypothesized relationships of these variables. Section 3describes the research methodology, while Section 4 reports the findings. The last section includes a discussion of the findings.

2. LITERATURE REVIEW

2.1. Individual knowledge creation behavior

According to the value creation perspective, knowledge creation is crucial in maintaining a business's new value creation, a point that has been verified by a number of scholars and studies (Felin and Hesterly, 2007 and Lepak et al., 2007). Yet are there not other factors that affect the subtle interplay of knowledge creation and competitiveness? Knowledge creation originates with individuals, and gradually accumulates in the minds of the members of an organization (tacit knowledge). Through systematic articulation, individuals make this personal knowledge available to the entire organization, thus turning tacit into explicit knowledge. The organization then collects and combines these scattered pieces of knowledge into an integrated whole, which serves as a pool of explicit information to all the organization's members, allowing them to increase and expand the boundaries of their tacit knowledge.

Lam (2000) asserts that individual knowledge creation is a part of an organization's knowledge creation pool, and that it comprises all the knowledge, information, and skills an individual possesses. Such knowledge creation can be employed for particular tasks, or the solving of specific problems, but since there are cognitive limitations on an individual's capacity for storing and processing information, individual knowledge creation often tends to be of a tacit nature. Meanwhile, Tsoukas and Vladimirou (2001)also stresses the important role played by individuals in innovative processes. Other scholars who place a firm focus on individual-level knowledge creation within organizations include Brown and Woodland (1999), Gupta and Govindarajan (1991) andNonaka (1994).

The creation of knowledge or ideas stands at the beginning of every individual creation behavior. Individual knowledge creation behavior involves continuous interaction with the external world (Nonaka, 1994). The know-how and information that individuals gain over time forms their knowledge stocks. Current knowledge stocks shape the scope and direction of the search for new knowledge, so knowledge creation is a path-dependent process. The term "individual knowledge creation behavior" here describes the behaviors by which an individual acquires new knowledge (i.e. knowledge previously not in the individual's possession). Possible sources for an individual's acquisition of new knowledge include: (1) individual research, experimentation, and learning/experience (e.g. learning by doing); (2) survey and investigation of external resources; (3) other systematic processes for the acquirement of new knowledge (MCFadyen & Cannella, 2004).

2.2. Knowledge incentive mechanisms

Approaching knowledge incentive mechanisms from the ownership angle,[1] this study find that knowledge incentive mechanisms involve the allocation of property rights (Jones, 1983), which basically means the allocation of incentives (Barzel, 1997). As this study have stated, knowledge incentive mechanisms are defined as mechanisms that a principal interposes between the principal and an agent/agents, allowing the principal to provide the agent(s) with incentives, and to supervise/control the agent's or agents' activities. Knowledge incentive mechanisms are knowledge governance mechanisms from a property rights perspective.[2] Motivation mechanisms in this context include searching for knowledge/information and knowledge sharing, as well as accumulation and enhancement of human capital (Foss & Mahnke, 2003). Property rights function as a powerful incentive for individual knowledge creation (Brown et al., 1999a and Brown et al., 1999b)—an insight that dovetails well with Vroom's (1964) expectancy theory of motivation, which states that stock sharing and profit sharing not only provide strong financial incentives on the individual level, but also help to improve a corporation's overall performance. Brown et al., 1999a and Brown et al., 1999b also observes that property rights sharing arrangements have a positive effect on a corporation's performance. Property rights sharing arrangements are used widely in businesses. Generally, property rights sharing arrangements in knowledge management context include property rights sharing arrangement of R&D and property rights sharing arrangement (Brandes et al., 2003 and Grandori, 2001). Let us now have a look at various knowledge incentive mechanisms with regard to property rights:

2.2.1. Property rights sharing arrangement of R&D (PRSA of R&D)

Property rights sharing arrangement of R&D (PRSA of R&D) in this context means that individual employees who participate in joint venture projects and R&D work are given the opportunity to reap part of the economic yield resulting from a given R&D project, mostly through a share in intellectual property rights. The sharing of intellectual property right thus becomes a crucial part of any collaborative project. Generally, the parties involved in such a project will determine the ratio of intellectual property right allocation in accordance with the relevant technology licensing contracts. With the help of mathematical formulae and expressions, researchers have arrived at the following theoretical conclusions about how R&D units or firms allocate property rights within incomplete contractual frameworks (Grandori, 2001): (1) when buyer or seller acquire the property rights to the entire results of an R&D project, or where all the property rights belong to one party, it becomes impossible to compensate for previous under-investment or eliminate opportunism; (2) the optimal allocation of R&D property rights is achieved when the party with the largest influence on the project's surplus is granted the biggest share of that surplus; (3) the establishment of R&D joint ventures serves as a way of encouraging the parties involved to invest in fields that are seminal for the development of society as a whole.

2.2.2. Property rights sharing arrangement (PRSA)

Property rights sharing arrangement (PRSA) is a mechanism under which individual employees are awarded an extra income based on the margin of the company's (or one of its department's) profits. In other words, it is a system under which staff participates in the enterprise's economic fortune via optional contracts and at a fixed ratio (Sherman, Bohlander, & Chruden, 1988). PRSA can thus be described as an added value or bonus that serves to motivate employees to actively contribute to the corporation's overall performance.

2.3. Knowledge incentive mechanisms and individual knowledge creation behavior

The more diverse existing knowledge is, the better new knowledge can be absorbed.Cohen and Levinthal (1990) explain how new knowledge enters a corporation via individuals, and how such new knowledge needs to be related to existing knowledge in some way. Individual members of an organization possessing a whole range of knowledge in different fields will be better equipped to recognize the importance of new knowledge and its relationship to existing information. They will also be able to link up the two types of knowledge, combining them in useful ways. Knowledge incentive mechanisms can manage knowledge resources through market competition, price, and contracts. An example for this is PRSA of R&D. Other ways for knowledge incentive mechanisms to manage knowledge resources include profit and stock sharing schemes. The value of property rights is directly connected to employer–employee relationships (Shperling, Rousseau, & Ferrante, 2003): increased participation in company affairs and stock ownership allows employees to assume a more active role, and encourages them to create new knowledge and thus new assets for the corporation.

Another aspect of this is that knowledge incentive mechanisms function as allocators of property rights (Barzel, 1997). This happens as a way of allocating motivation, including motivation for the creation of new knowledge (Foss & Mahnke, 2003). Particularly within knowledge incentive mechanisms, high-powered performance incentives may adequately strengthen individual innovate behavior within an organization (Teece, 2003). High return knowledge creation is almost always individual knowledge creation, because the prospect of profits produces the strongest results on the individual level. Put differently, the level of income among a corporation's employees will certainly have a palpable influence on these employees' motivation to actively engage in innovative behavior. From all the above, this study can formulate the influence of knowledge incentive mechanisms on individual creation behavior.

2.4. Knowledge psychological ownership

Pierce et al. (1991) believes that there are two levels on which ownership affects employees' behavior. The first level is ownership, but the second level involves psychological ownership. Ownership is created through practical arrangements within the corporate system, but psychological ownership has to be carefully fostered through multifaceted socio-psychological effects and factors that can influence employee behavior (Van Dyne & Pierce, 2004). In other words, genuine ownership is the result of psychological ownership, and employees need to feel secure of the latter if the company wants to steer their behavior and performance in a positive direction. At the same time, physical ownership has a positive effect on psychological ownership.

Psychological ownership of knowledge is defined as a mindset in which employees perceive that a certain knowledge or information belongs to them. Within an organization, it is vital that employees develop this kind of perception or awareness, and feel closely connected with the knowledge they own (Dittmar, 1992), because only through this kind of psychological link can knowledge become a part of the individual's identity (Pierce et al., 1991). As this study have stated, people are motivated to strive for the possession of knowledge because it gives them control, and as an individual acquires ownership and thus control, he or she experiences efficacy and effectance. These basic experiences are then utilized to influence individual perception and identification, a process that prompts individuals to set in motion, control, and maintain the kinds of behavior expected from them by the organization (Stajkovic & Luthans, 1998). Let us have a closer look at this psychological process: before people begin to engage in innovative behavior, they tend to carefully consider and evaluate their situation, and try to estimate whether or not they are currently in possession and control of knowledge. Individual expectance of efficacy will determine whether or not, and to what extent, the individual will engage in, and maintain, the kind of behavior that the organization expects from it. If the individual feels that it does indeed own relevant knowledge, and considers itself capable of high efficacy and efficiency, it will then devote its entire efforts to completing the tasks or duties that the organization has given to it. On the other hand, if the individual perceives that it does not possess any relevant knowledge, and deems its own efficacy/efficiency to be low, it is quite likely that it will abandon its efforts prematurely, and thus cause a task or mission to fail.

2.5. Knowledge incentive mechanisms and knowledge psychological ownership

Knowledge incentive mechanisms endow employees with shared physical ownership in specific objects and/or their financial value, allowing them to enjoy the benefits of control over specific property (Pierce et al., 1991). Organizations employ knowledge incentive mechanisms (PRSA of R&D and PRSA) to influence the function and role of employees, granting employees' participation in ownership and letting them develop a sense of psychological ownership of knowledge. The latter is crucial in enabling employees to cultivate a sense of physical ownership, which in turn will make them much more likely to behave and perform in the way the organization desires. In addition, the more property rights individuals are given, the more likely they are to develop a sense of psychological ownership of knowledge (Pierce et al., 1991). When employees are provided with physical ownership (e.g. in the form of company shares), they will experience feelings of success and other positive emotions, which will further contribute to their sense of psychological ownership of knowledge. Reviewing the literature on psychological ownership, Furby (1978) finds that when people develop a sense of ownership towards a certain target, they will begin to view that target as an extension of their own person. This will lead to the generation of a self-concept, an image projected towards the environment (Weick, 1995). As Hall (1996) points out, awareness of psychological ownership makes people want to protect or defend their property rights. This research discovered that the higher the employees' sense of psychological ownership of knowledge, the more likely the employees are to view this knowledge as their exclusive property, and this will in turn lead to more egoistic forms of protective or defensive behavior. From these findings about perceived exclusive ownership of knowledge, this study can deduce the next two hypotheses:

Hypothesis 1.

The more aware individuals are of PRSA of R&D (knowledge incentive mechanisms) within their organization, the higher their sense of psychological ownership of knowledge will be.

Hypothesis 2.

The more aware individuals are of PRSA (knowledge incentive mechanisms) within their organization, the higher their sense of psychological ownership of knowledge will be.

2.6. Knowledge psychological ownership and individual knowledge creation behavior

This research confirms that sense of knowledge psychological ownership is an important factor for predicting the behavior of employees (Pierce et al., 1991, Pierce et al., 2003,Pierce et al., 2003, Van Dyne and Pierce, 2004 and VandeWalle et al., 1995). When employees do have a sense of psychological ownership of knowledge, this may lead to three possible outcomes: (1) it will have a positive effect on their innovative behavior/willingness to create new knowledge; (2) the employees will view their knowledge as extension of themselves; (3) they will develop a sense of responsibility towards the corporation, perceiving it as their duty to create new knowledge.

Employees that have a sense of psychological ownership of knowledge feel that they possess this knowledge (Dittmar, 1992), and develop a close psychological link towards this knowledge (Pierce et al., 1991). In other words, the knowledge becomes an integral part of their individual identity. When individuals perceive in this way that knowledge is theirs, they define their own position within the organization in terms of their control over knowledge. This means they use ownership of knowledge as a yardstick of their own value, and judge their own performance by the knowledge they have. Individuals tend to gain a sense of achievement from their control of knowledge, which becomes an important factor of their internal self-evaluation process (Huber, 1991). For the purposes of this research, an employee's innovative behavior is defined as individual acquisition and/or generation of new knowledge (i.e. knowledge not previously in the individual's possession). If during or as a result of this process an individual develops a relatively keen sense of ownership of knowledge, the individual is then more likely to engage in innovative behavior more or less exclusively aimed at maximizing its own interests. When this happens in a corporate scenario, employees tend to focus on behavior that helps them to create and accumulate new knowledge, but are reluctant to engage in sharing behavior would run counter to their own perceived best interests. Therefore, this study arrives at the third hypothesis. The hypothesized model is shown in Fig. 1.

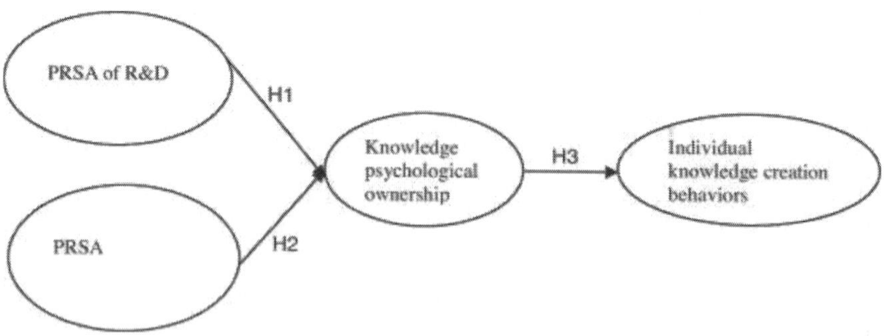

Figure 1. Hypothesized model.

Hypothesis 3.

The higher a knowledge user's sense of psychological ownership of knowledge, the more likely he is to engage in individual knowledge creation behavior.

3. METHODOLOGY

3.1. Sampling and respondents

Taiwanese high-tech industry offers an excellent context within which knowledge incentive mechanisms promote employees' knowledge creation behavior. In the early 1980s, Taiwanese high-tech industry started to develop in the form of original equipment manufacturers. Afterward, to enhance its competitiveness and profitability, Taiwanese high-tech industry gradually upgraded itself by engaging in product design and production management (Tsai & Hung, 2006). Now, Taiwanese high-tech businesses are capable of undertaking product design, development and manufacturing but are not directly engaged in marketing and brand promotion. The reason why Taiwanese high-tech businesses can be studied are the followings: first, Taiwan has high quality R&D talents as well as the advantage of relatively low cost of R&D. Second, the R&D outsourcing of US companies, coupled with the US companies" focusing their efforts on marketing and brands, also offers Taiwanese high-tech industries an opportunity to transform and promote themselves (Liu and Liu, 2011 and Tsai and Hung, 2006).

In addition, R&D professionals are widely acknowledged as knowledge workers. Their works are characterized by complex system design, knowledge related to the rapid update of technology, and strong competition for sustaining innovation (Assimakopoulos & Yan, 2006). In addition, the knowledge creation behaviors among R&D professionals are quite different from those of traditional workers (Berends et al., 2006, Cabrera and Cabrera, 2002 and Liu and Liu, 2008). Since the R&D process is extremely complicated and an employee only possesses expertise in a specific area, knowledge will not be exposed to other individuals until knowledge owners make the objects available.

Table 1. Respondent demographic information.

Variable	Frequency	Percentage (%)
Gender		
Male	251	60.9
Female	161	39.1
Age		
Under 30	169	41.0
30–39	196	47.6
40–49	41	10.0
Over 50	6	1.4
Education		
Bachelor's degree	360	87.5
Graduate degree	92	22.3
Ph.D. degree	1	0.2
Tenure in present jobs		
Under 5	243	59.0
6–9.9	125	30.3
10–14.9	29	7.1
15–19.9	10	2.4
Over 20	5	1.2
Manager		
Yes	60	14.6
No	352	85.4

Therefore, it is valuable to investigate how R&D professionals create knowledge among members. Since knowledge creation behaviors are so critical for R&D professionals, the current sample consists of R&D employees employed in high-tech companies in Taiwan. The companies studied are engaged in applied research within the computer systems and other industries, which are pivotal for high-tech industries.

The research subjects for this study were expatriates from Taiwanese enterprises operating abroad. Concerning the difficulties of choosing appropriate samples for our study, this study used convenient sampling to collect data. In Asian countries, personal relationships are very helpful in collecting research data. Our team members used their personal relationships to contact potential samples and sent questionnaires to R&D employees who had indicated their willingness to participate in this survey. Data collection was conducted during the first quarter of 2010. The survey was processed through multi-channels, including visits in person, regular mail, e-mail, and web sites. A preliminary test showed that the sources of data collection make no significant difference in results. In all, 460 questionnaires were completed. Of these, 48 were rejected on the basis of missing data, leaving a useable sample size of 412 for analysis. The respondent demographic information is shown in Table 1.

3.2. Measures

3.2.1. Independent variables

This study used eight items based on the theoretical literature to assess possible knowledge incentive mechanisms in high-tech industries. The items were rated on a 6-point Likert scale. The eight items used to assess knowledge incentive mechanisms were based on the discussion of knowledge governance mechanisms by Grandori (2001),[3]Bai, Tao, & Wu (2004)Bai, Tao, and Wu (2004) and Brown et al., 1999a and Brown et al., 1999b, including discussions of *property rights sharing arrangement of R&D* and *property rights sharing arrangement*. Property rights sharing arrangement included stock sharing, profit sharing and other incentives (Bai et al., 2004,Brown et al., 1999a and Brown et al., 1999b).

3.2.2. Mediator variables

This study used seven items based on the theoretical literature to assess the possible knowledge psychological ownership among R&D professionals. The items were rated on a 6-point Likert scale regarding knowledge psychological ownership among R&D professionals. The seven items used to assess knowledge psychological ownership were based on the discussion of psychological ownership toward the organization by Van Dyne and Pierce (2004), but this study reworded it to test participants' feelings of possession towards their knowledge (e.g. "This is MY knowledge"[4])(Mayhew et al., 2007 and Pierce et al., 2004). These items involved a reverse-worded item.

3.2.3. Dependent variables

Individual knowledge creation behaviors were the major dependent variables of interest in this study. Knowledge creation among R&D professionals, which was defined as the process by which an individual acquires new knowledge, was rated by a 6-point Likert scale and measured by three questions. The *individual knowledge creation behaviors*include "individual research, experimentation, and learning experience", "survey and investigation of external resources" and "other systematic processes for the acquirement of new knowledge". The measures of this variable were adapted from the theoretical literature (Liu, 2008).

Confirmatory factor analysis (CFA) was used to examine the validity and reliability of the research instrument. Validity was addressed in two ways. First, operationalization of the constructs drew upon an extensive literature review, and questionnaires were discussed with superiors in actual organizations. Both processes were intended to enhance content validity. Second, the results of the CFA illustrated that four variables displayed convergent validity, since the analysis yielded four factors with factor loading displaying the expected patterns. Table 2 shows the correlations among the four factors of the CFA model. The internal-consistency coefficients of the reliability of the unity are weighted sums of the item scores, and the α-coefficients vary between 0.86 and 0.94. The Cronbach's α coefficients of the four variables of this study all exceeded the 0.70 threshold, with the reliability and internal consistency of the variables confirmed.

Table 2. Correlations, average variance extracted (AVE) and the square root of average variance extracted among the factors of the measurement model.

Variable	AVE	1	2	3	4
1. PRSA of R&D	0.68	**0.82**			
2. PRSA	0.84	0.72	**0.92**		
3. Knowledge psychological ownership	0.84	0.37	0.35	**0.92**	
4. Individual knowledge creation behaviors	0.71	0.25	0.38	0.18	**0.84**

$N = 412$.
AVE is average variance extracted.
The diagonal of the matrix presents the square root of average variance extracted.

The diagonal of the matrix presents the square root of average variance extracted.

3.2.4. Construct validity

In general, construct validity refers to the degree to which instrument truly measure the constructs for what it is intended to measure. There are several types of construct validity measures that include the convergent and discriminating validity. Convergent validity is measured by average variance extracted (AVE) for each construct during the reliability analysis that should be 0.5% or 50% or better (Fornell & Larcker, 1981). Table 2 shows the average variance extracted (AVE) for the constructs ranged from 0.68 to 0.84 indicating a sufficient level of convergent validity of all constructs. To further verify discriminating validity, Fornell and Larcker (1981) advocate that correlations between items in any two constructs should be lower than the square root of the average variance extracted shared by items within a construct. As shown in Table 2, the square root of the average variance extracted shared between a construct and its items was greater than the correlations between the construct and any other construct in the model, satisfyingFornell and Larker's (1981) criteria for discriminating validity. The above results, therefore, confirm that our instrument encompassed satisfactory construct validity.

3.2.5. Common method variance

The data for the independent, mediator variables and dependent variables were obtained from the same respondents, which may have introduced a common method bias. This study controlled for this bias both methodologically and statistically. First, this study took steps to protect respondents' anonymity and counterbalanced the order of items. Second, this study employed Harman's single-factor test (Podsakoff, MacKenzie, Lee, & Podsakoff, 2003). Specifically, this study incorporated all the variables in an exploratory factor analysis and examined the unrotated factor solution to determine the number of factors necessary to account for the variance. The results revealed no general factor (the loading for the first principle component was only

0.35), suggesting that there is no systematic variance common to the measures.

4. RESULTS

4.1. The measurement model

This study tested the factor structure of our survey measure using confirmatory factor analysis (CFA). The CFA model was developed with the purposes of testing whether the hypothesized latent variables could be identified empirically, and for examining the validity and reliability of the measures. First, according to the successive testing and modification of different models on the same data of course, different variable sets were combined into the final model. The final CFA model had a χ^2 value of 96.36, $df = 48$ ($\chi^2/df = 2$) ($p = 0.00$). Other indices of fit also showed acceptable fit (GFI = 0.96, CFI = 0.99, NFI = 0.98, NNFI = 0.98). Standardized factor loadings, along with descriptive item data, are shown in Table 3.

Table 3. Means, standard deviations, and factor loadings for the items.

Item	Item no.	Mean	SD	Standard factor loading
PRSA of R&D				
Rate of the stock sharing of R&D	PR1	3.13	1.46	0.69
Rate of the profit sharing of R&D	PR2	2.59	1.46	0.86
Incentives for R&D project	PR3	2.73	1.53	0.90
PRSA				
Rate of the stock sharing	PS1	3.00	1.56	0.83
Rate of the profit sharing	PS2	2.69	1.53	0.97
Rate of variable pay	PS3	2.78	1.56	0.95
Knowledge psychological ownership				
This is MY knowledge	KP1	3.50	1.37	0.86
I feel a very high degree of personal ownership for knowledge	KP2	3.50	1.35	0.98
I sense that this is MY knowledge	KP3	3.53	1.34	0.91
Individual knowledge creation behaviors				
Individual research, experimentation, and learning experience	CB1	4.28	1.00	0.70
Survey and investigation of external resources	CB2	4.31	1.05	0.87
Other systematic processes for the acquirement of new knowledge	CB3	4.19	1.10	0.94

Eight items measured the *knowledge incentive mechanisms*. After a principal component analysis, this study removed items PR4 and PR5, loading by less than 0.5. Together, the two factors were able to explain over 79.57% of the total variation in the six items. Bartlett's test of sphericity was significant ($p < 0.0001$). The Kaiser–Meyer–Olkin statistic was 0.73, suggesting that individual measures of sampling adequacy were generally high. Three items measured the *PRSA of R&D*. Three items with loadings from 0.69 to 0.930 asked about “the rate of the stock sharing of R&D”, “the rate of the profit sharing of R&D” and “the incentives for R&D project”. Cronbach's α for the full 3-item scale is 0.86. The *PRSA* factor was related to three items. The highest loading (0.97) was observed for the item that asked about “the rate of the profit sharing”. Another loading (0.95) was observed for the item that asked about “the rate of variable pay”. The lower loading (0.83) asked about “the rate of the stock sharing”. Cronbach's α for the full 3-item scale is 0.94.

Seven items measured the *knowledge psychological ownership*. After a principal component analysis, this study removed items KP 4, KP 5 and KP6, loading by less than 0.5. The item KP7 (reverse-worded item) was dropped. Together, the factor was able to explain over 81.71% of the total variation in the three items. Bartlett's test of sphericity was significant ($p < 0.0001$). The Kaiser–Meyer–Olkin statistic was 0.76, suggesting that individual measures of sampling adequacy were generally high. Three items measured the *knowledge psychological ownership* factor, and all items had substantial loadings (0.86–0.98) on the factor. The highest loading (0.98) was observed for the item that asked about “I feel a very high degree of personal ownership for knowledge”. Another loading (0.91) was observed for the item that asked about “I sense that this is MY knowledge”. The lower loading (0.86) asked about “this is MY knowledge”. Cronbach's $\alpha$for the full 3-item scale is 0.94.

Three items measured the i*ndividual knowledge creation behaviors* factor. After a principal component analysis, the factor was able to explain over 79.82% of the total variation in the three items. Bartlett's test of sphericity was significant ($p < 0.0001$). The Kaiser–Meyer–Olkin statistic was 0.72, suggesting that individual measures of sampling adequacy were generally high. Three items measured the i*ndividual knowledge creation behaviors* factor, and all items had substantial loadings (0.70–0.94) on the factor. The highest loading (0.94) was observed for the item that asked about “other systematic processes for the acquirement of new knowledge”. Another loading (0.87) was observed for the item that asked about "survey and investigation of external resources". The lower loading (0.70) asked about "individual research, experimentation, and learning experience". Cronbach's α for the full 3-item scale is 0.87.

4.2. The structural model

In the next step of modeling, path models were fitted to test the proposed model. The hypotheses do not explicitly specify the relations among all the factors, and some unexpected relations were found, so a sequence of models was fitted. The selection criteria for the final model were: (1) fit to data and (2) interpretability of the estimated relations.

The path model had a test statistic of $\chi^2 = 151.28$, $df = 50$ ($\chi^2/df = 3^5$) ($p = 0.00$). Other indices of fit also showed the fit to be acceptable (GFI = 0.94, CFI = 0.98, NFI = 0.96, NNFI = 0.97), so this study cannot reject the hypothesis that the path model correctly reproduced the correlations among the latent variables. Fig. 2 shows the model.

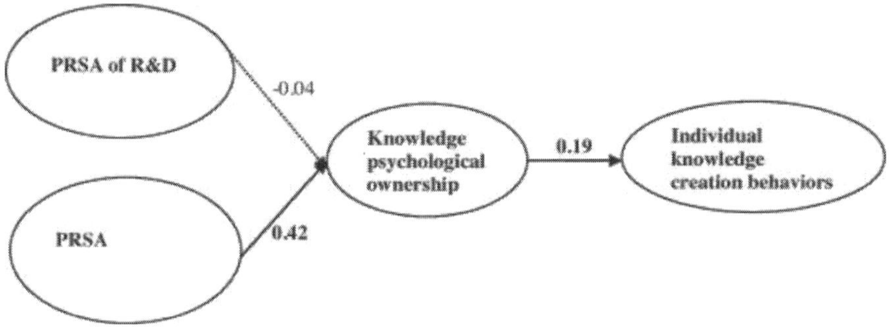

Figure. 2. Path model for relations among the latent variables. *Note*: Path coefficients are standardized partial regression coefficients estimated by maximum likelihood. Ellipses represent latent variables. Unidirectional arrows express direct effects. *T*-values above 1.96 ($p < 0.05$) are shown in bold. $N = 412$.

The pattern of direct effects revealed by the path model seems to provide somewhat mixed evidence for the study's hypotheses.

According to Hypothesis 1, this study expected a positive relationship between PRSA of R&D and knowledge psychological ownership. However, the path model reveals no significant correlations between PRSA of R&D and knowledge psychological ownership (−0.04). This factor likewise failed to demonstrate significant correlations in the model. The reason could be Taiwanese high-tech industry gradually upgraded itself from original equipment manufacturers (OEM) to original product design manufacturers (ODM) (Tsai & Hung, 2006). After the mid-1990s, Taiwanese manufacturers took a step further to stress creativity and design capacity. Because shifting upward from OEM to ODM was in a short period of time, R&D professionals could not distinguish PRSA of R&D from PRSA. Thus, the path model reveals no significant correlations between PRSA of R&D and knowledge psychological ownership.

According to Hypothesis 2, this study expected a positive effect from PRSA on knowledge psychological ownership. The result of the path model supported this hypothesis: there was a positive direct effect of PRSA on knowledge psychological ownership (0.42). Hypothesis 3 proposed that there should be a positive effect from the knowledge psychological ownership on individual knowledge creation behaviors. The path model showed a positive effect of the knowledge psychological ownership on individual knowledge creation behaviors (0.19). One of knowledge incentive mechanisms, PRSA, influences individual knowledge creation behaviors via knowledge psychological ownership. Thus, the results of the path model support hypotheses of this research.

5. DISCUSSION

A contribution of this study is our research on interventions – the knowledge creative link was our conclusion from the integrated model, in which knowledge incentive mechanisms including PRSA was found to have a positive effect on knowledge psychological ownership, and to thereby influence individual knowledge creation behaviors among R&D professionals.

Our integrated model of knowledge creative behavior leads us to expect that the "knowledge psychological ownership" will mediate the relationship between knowledge incentive mechanisms and individual knowledge creation behaviors among R&D professionals. Path analysis suggests that the knowledge psychological ownership may mediate "PRSA" and "individual knowledge creation behaviors". This study found that the mediation models existed in our integrated knowledge creative model. "Knowledge psychological ownership" mediated the relationship between "PRSA" and "individual knowledge creation behaviors". This implies that the factor of "knowledge psychological ownership" plays a very important role in the integrated knowledge creative model. It may be that the characteristics of the knowledge used by R&D professionals in high-tech industries – complex system design knowledge, rapid technology upgrades, and constant innovation – are necessary to maintain competitiveness (Assimakopoulos & Yan, 2006). As discussed earlier, if R&D professionals feel that they do indeed own relevant knowledge, and considers themselves capable of high efficacy and efficiency, they will then devote their entire efforts to completing the tasks or duties that the organization has given to them.

Therefore, findings from evidential analyses include: (1) A new assessment of knowledge incentive mechanisms was helpful in gaining a more detailed understanding of individual knowledge creation behavior. (2) The psychological ownership of knowledge can encourage individual knowledge creation behavior. (3) Knowledge incentive mechanisms stress private and individual proprietary characteristics employees who possess real ownership (i.e. corporate stock) when they innovate will deem such innovation as an extension of self, giving rise to a protective or defensive mindset, which then has an impact on individual knowledge creation behaviors because of the psychological ownership of knowledge.

6. IMPLICATIONS

6.1. Theoretical implications

This research's contribution to KM and related fields consists of establishing a comprehensive multi-construct analytical framework for the examination of individual KM. This research applies the theory of governance mechanisms (which is a part of organizational theory) and the theory of psychological ownership (which is an important concept in the field of organizational behavior) to the study of individual innovative behavior (i.e. creation of new knowledge). The goal is to explain how organizations use knowledge incentive mechanisms to influence individual knowledge creation behavior through the effects of (the sense of) psychological ownership of knowledge. In the following this study briefly discussed the concrete contributions this research makes to the existing theories in various relevant fields.

6.1.1. Examining knowledge incentive mechanisms from the perspective of ownership

This research breaks with the traditional approach of exploring the meaning and function of knowledge incentive mechanisms by looking at hierarchy-based, price/contract-based, and identity-based forms of governance, and attempts instead to explore the relevant issues from the angle of ownership. More specifically, this study examines knowledge incentive mechanisms and through application of theories of psychological ownership, to find out which factors have a bearing on management processes that aim at encouraging individual innovative behavior. This research makes two specific contributions to the theory of knowledge incentive mechanisms: (1) approaching the topic from an ownership perspective, this study proposes a new categorization of knowledge incentive mechanisms that is conducive to gaining more detailed insights into patterns of individual knowledge behavior; (2) employing ownership theory, this study merges aspects of organizational systems (knowledge incentive mechanisms) and individual psychological states (e.g. psychological ownership of knowledge) for an integrated view that allows a more exhaustive and precise description of the management processes that govern individual innovative behavior within organizations.

6.1.2. Methods of encouraging creation of new knowledge

Through the analysis of the relevant literature, this research discovered that individual creation behavior follows certain patterns that are most accurately expressed in the theories of knowledge incentive mechanisms and the creation of new knowledge. Knowledge incentive mechanisms emphasize personal gain and individual ownership: by creating new knowledge, employees may obtain physical ownership (e.g. in the form of shares), and as a result they will view any new knowledge they create as an extension of their own person (Furby, 1978), which in turn generates a protective or defensive attitude (Hall, 1996). This sense of psychological ownership of knowledge provides leverage for influencing individual creation behavior.

6.1.3. Applying the theory of psychological ownership to the specific sense of psychological ownership of knowledge

Making use of the psychometric theory of psychological ownership, this research employed questionnaires based on rigorous psychometric scales and procedures. This study formulated questions and parameters for assessing the potential targets of psychological ownership. Our goal was to expand the scope of assessment from the original organizational patterns to knowledge itself. This helped us to ensure the general applicability of our questionnaire results, as well as allowing for a proper representation of the issue's complexity. In this study echo the views of Van Dyne and Pierce (2004), who stated that the idea of psychological ownership may well be applied to the assessment of problems outside of organizational theory (e.g. equipment, technology, products and service procedures that form part of the working environment). Therefore, this study has also made a contribution to strengthening the positivistic basis of general constructs for measuring the sense of psychological ownership.

6.2. Practical implications

6.2.1. Increasing R&D professionals' perceived psychological ownership

With this study focus on the psychological processes that underlie knowledge creation, our integrated model allows us to link individuals' psychological ownership to their engagement in knowledge creation. The important role of individuals' psychological ownership on knowledge creation has also been highlighted in this study. The fact that the new knowledge R&D engineers often acquire is complex makes it more difficult to share. Thus, R&D professionals' perceived psychological ownership in knowledge creation is central to its occurrence, because actions are based more on what people believe they can do than on what they actually can do. Therefore, managers should strive to increase R&D professionals' perceived psychological ownership by establishing mechanisms through which employees receive feedback whenever knowledge creation takes place.

6.2.2. Formal knowledge incentive mechanisms being beneficial for the promotion of knowledge creation behavior among R&D professionals

This study suggests that formal knowledge incentive mechanisms are beneficial for the promotion of knowledge creation behavior among R&D professionals. Though other researchers have suggested that firms need to create specific interventions for knowledge creation purposes, this study provides evidence that the regular organizational mechanisms are also important for shaping individuals' knowledge creation. The evidence from this study supports the idea that some knowledge incentive mechanisms may enhance employees' perceived psychological ownership to create knowledge. Knowledge management is, in fact, multidisciplinary. A firm needs to integrate its formal practices with specific knowledge-management initiatives to allow the creation of new knowledge. If firms lack sufficient formal practices and structural mechanisms, they may encounter a situation in which knowledge management initiatives do not add value.

7. LIMITATIONS

This research was subject to the following limitations. First, the sample used in this study consisted of large enterprises (with annual operating revenues in excess of US $3 billion). Future research may consider using data from SMEs. Second, the subjective assessment of knowledge creation behavior may not objectively express real knowledge creative performance. Future research should simultaneously employ subjective and objective assessment standards. This research may, for example, simultaneously evaluate self-assessment scores and R&D patent results.

ACKNOWLEDGMENT

This paper were founded by the National Science Council, Taiwan (project no.: NSC 99-2410-H-031-003- and NSC 100-2410-H-031-054-SSS).

REFERENCES

1. Assimakopoulos and Yan, 2006 D. Assimakopoulos, J. Yan **Source of knowledge acquisition for Chinese software engineers** R&D Management, 31 (1) (2006), pp. 97–105

2. Bai, C. E., Tao, Z., & Wu, C. (2004). Revenue sharing and control rights in team production: Theories and evidence from joint ventures. Rand Journal of Economics, 35(2), 277–305.

3. Barzel, Y. (1997). Economic analysis of property rights. Cambridge: Cambridge University Press Berends, H., Bij, H., Debackere, K., & Weggeman, M. (2006). Knowledge sharing mechanism in industrial research? R&D Management, 31(1), 85–95.

4. Brandes, P., Dharwadkai, R., & Lemesis, G. V. (2003). Effective employee stock option design: Reconciling stakeholder,

strategic, and motivational factors. Academy of Management Executive, 17(1), 77–92.

5. Bridoux, F., Coeurderoy, R., & Durand, R. (2011). Heterogeneous motives and the collective creation of value. Academy of Management Review, 36(4), 711–730.

6. Brown, R. B., & Woodland, M. J. (1999). Managing knowledge wisely: A case study in organizational behavior. Journal of Applied Management Studies, 8(2), 175–198.

7. Brown, S., Fakhfakh, F., & Sessions, J. G. (1999). Absenteeism and employee sharing: An empirical analysis based on French panel data, 1981–1991. Industrial and Labor Relations Review, 52(2), 234–251

8. Brown, S., Fakhfakh, F., & Session, J. (1999). Absenteeism and employee profit sharing: An empirical analysis based on French panel data, 1981–1991. Industrial and Labor Relations Review, 52, 234–251.

9. Cabrera, A., & Cabrera, E. F. (2002). Knowledge-sharing dilemmas? Organization Studies, 23(5), 687–710.

10. Child, J., & McGrath, R. (2001). Organizations unfettered: Organizational form in an information intensive economy. Academy of Management Journal, 44, 1135–1148.

11. Coff, R. W. (2003). The emergent knowledge-based theory of competitive advantage: An evolutionary approach to integrating economics and management? Managerial & Decision Economics, 24(4), 245–251.

12. Cohen, W. M., & Levinthal, D. A. (1990). Absorptive capacity: A new perspective on learning and innovation. Administrative Science Quarterly, 35, 128–152.

13. Dittmar, H. (1992). The social psychology of material possessions: To have is to be. New York: St. Martin Press. Fehr,

E., & Fischbacher, U. (2002). Why social preferences matter—The impact of nonselfish motives on competition, cooperation and incentives. Economic Journal, 112, 1–33.

14. Felin, T., & Hesterly, W. S. (2007). The knowledge-based view, nested heterogeneity, and new value creation: Philosophical considerations on the locus of knowledge. Academy of Management Review, 32(1), 195–218.

15. Fey, C. F., & Birkinshaw, J. (2005). External sources of knowledge, governance mode, and R&D performance. Journal of Management, 31(4), 597–621.

16. Fornell, C., & Larcker, D. (1981). Structural equation models with unobservable variables and measurement error. Journal of Marketing Research, 18(1), 39–50.

17. Foss, N. J. (2006). Knowledge and organization in the theory of the multinational corporation: Some foundational issues. Journal of Management and Governance, 10(1), 3–20.

18. Foss, N. J., & Mahnke, V. (2003). Knowledge management: What does organizational economics contribute? In M. Easterby-Smith, & M. Lyles (Eds.), Handbook of knowledge management. Oxford: Basil Blackwell.

19. Furby, L. (1978). Possession in humans: An exploratory study of its meaning and motivation. Social Behavior and Personality, 6, 49–65.

20. Grandori, A. (2001). Neither hierarchy nor identity: Knowledge-governance mechanisms and the theory of the firm. Journal of Management and Governance, 5, 381–399.

21. Gottschalg, O., & Zollo, M. (2007). Interest alignment and competitive advantage. Academy of Management Review, 32, 418–437.

22. Gupta, A. K., & Govindarajan, V. (1991). Knowledge flow and the structure of control within multinational corporations. Academy of Management Review, 16(4), 768–792.

23. Hall, E. T. (1996). The hidden dimension. Garden City, New York: Doubleday. Hsu. (2006). Enhancing employee tendencies to share knowledge-case studies of nine companies in Taiwan. International Journal of Information Management, 26(4), 326–338.

24. Huber, G. P. (1991). Organizational learning: The contributing processes and the literatures. Organization Science, 2, 88–115.

25. Insead, R. K., & Lim, K. (2007). The impact of acquisitions on the productivity of inventors at semiconductor firms: A synthesis of knowledge-based and incentive-based perspectives. Academy of Management Journal, 50(5), 1133–1155.

26. Jones, G. R. (1983). Transaction costs, property rights, and organizational culture: An exchange perspective. Administrative Science Quarterly, 28, 454–467.

27. Lam, A. (2000). Tacit knowledge, organizational learning and societal institutions: An integrated framework. Organization Studies, 21(3), 487–513.

28. Lepak, D. P., Smith, K. G., & Taylor, M. S. (2007). Value creation and value capture: A multilevel perspective. Academy of Management Review, 32(1), 180–194.

29. Liu, M. (2008). Impact of knowledge governance mechanisms on individual knowledge management behavior–An ownership perspective discussion. Ph.D. dissertation, National Central University.

30. Liu, M., & Liu, N. (2008). Sources of knowledge acquisition and patterns of knowledge sharing behaviors—An empirical study of

Taiwanese high-tech firms. International Journal of Information Management, 28(5), 423–432.

31. Liu, N., & Liu, M. (2011). Human resources practices and individual knowledge sharing behavior: An empirical study for Taiwanese R&D professional. International Journal of Human Resource Management, 22(4), 981–997.

32. Mayhew, M. G., Ashkanasy, N. M., Bramble, T., & Gardner, J. (2007). A study of the antecedents and consequences of psychological ownership in organizational settings. Journal of Social Psychology, 147(5), 477–500.

33. MCFadyen, M. A., & Cannella, A. A. (2004). Social capital and knowledge creation: Diminishing returns of the number and strength of exchange relationships. Academy of Management Journal, 47(5), 735–746.

34. Nickerson, J., & Zenger, T. (2004). A knowledge-based theory of the firm: The problem-solving perspective. Organization Science, 15(6), 617–632.

35. Nickerson, J., & Zenger, T. (2008). Envy, comparison costs, and the economic theory of the firm. Strategic Management Journal, 29, 1429–1449.

36. Nonaka, I. (1994). A dynamic theory of organizational knowledge creation. Organization Science, 5, 14–37.

37. Osterloh, M., & Frey, B. (2000). Motivation, knowledge transfer and organizational form. Organization Science, 11, 538–550.

38. Ostrom, E. (2000). Collective action and the evolution of social norms. Journal of Economic Perspectives, 14(3), 137–158.

39. Parker, S. K., Wall, T. D., & Jackson, P. R. (1997). That's not my job: Developing flexible employee work orientations. The Academy of Management Journal, 40(4), 899–929.

40. Pierce, J. L., Kostova, T. L., & Dirks, T. K. (2003). The state of psychological ownership: Integrating and extending a century of research. Review of General Psychology, 7, 84–107.

41. Pierce, J. L., O'Driscoll, M. P., & Coghlan, A. M. (2004). Work environment structure and psychological ownership: The mediating effects of control. Journal of Social Psychology, 144(5), 507–534.

42. Pierce, J. L., Rubenfeld, S. A., & Morgan, S. (1991). Employee ownership: A conceptual model of process and effects. Academy of Management Review, 16, 121–144.

43. Podsakoff, P. M., MacKenzie, S. B., Lee, J. Y., & Podsakoff, N. P. (2003). Common method biases in behavioral research: A critical review of the literature and recommended remedies. Journal of Applied Psychology, 88(5), 492–499.

44. Sherman, A. W., Bohlander, G. W., & Chruden, H. J. (1988). Managing human resources. NY: John Wiley & Son.

45. Shperling, Z., Rousseau, D. M., & Ferrante, C. J. (2003). Ownership in high technology firms: Israel and U.S. technical report. Pittsburgh: Carnegie Mellon University, Heinz School of Public Policy and Management.

46. Stajkovic, A. D., & Luthans, F. (1998). Self-efficacy and work-related performance: A meta-analysis. Psychological bulletin, 124(2), 240–261.

47. Teece, D. J. (2003). Expert talent and the design of (professional services) firms. Industrial & Corporate Change, 12, 895–916.

48. Tsai, J., & Hung, H. (2006). The competitive strategies of Taiwanese hi-tech industry? Journal of Global Business Management, 2(2), 118–125.

49. Tsoukas, H., & Vladimirou, E. (2001). What is organizational knowledge? Journal of Management Studies, 38, 973–993.

50. Van Dyne, L., & Pierce, J. L. (2004). Psychological ownership and feelings of possession: Three field studies predicting employee attitudes and organizational citizenship behavior. Journal of Organizational Behavior, 25(4), 439–459.

51. VandeWalle, D., Van Dyne, L., & Kostova, T. (1995). Psychological ownership: An empirical examination of its consequences. Group and Organization Management, 20(2), 724–749.

52. Vroom, V. (1964). Work and motivation. New York: McGraw-Hill. Weick, K. E. (1995). Sense making in organizations. Thousand Oaks, CA: Sage Publications, Inc.

53. Ziedonis, R. H. (2004). Don't fence me in: Fragmented markets for technology and the patent acquisition strategies of firms. Management Science, 50, 804–820.

Chapter 11

Integrated Access to Hybrid Information Resources

Tamar Sadeh

ABSTRACT

This paper is entitled "Integrated access to hybrid information resources", but integrated access is just a means to reach an end. Our challenge is to create an integrated environment in a heterogeneous world. Users are not aware, and do not want to be aware, of the differences between the various information resources that their institution provides. All they want is to be able to attain the information they are looking for, in the simplest and most straightforward way possible.

SIMULTANEOUS SEARCHING

Simultaneous searching refers to a process in which a user submits a query to numerous information resources. The resources can be heterogeneous in many aspects: they can reside in various places, offer information in various formats, draw on various technologies, hold various types of materials, and more. The user's query is broadcast to each resource, and results are returned to the user. The development of software products that offer such simultaneous searching relies on the fact that each information resource has its own search engine. The simultaneous searching product transmits the user's query to that search engine and directs it to perform the actual search. When the simultaneous searching software receives the results of the search, it displays them to the user.

Simultaneous searching is also known as *integrated searching, metasearching, cross-database searching, parallel searching, broadcast searching,* and *federated searching.* [1]MetaLib, the library portal from Ex Libris, provides such simultaneous searching with its Universal Gateway component. In this paper, we shall refer to these systems as metasearch systems.

Let's take a look at an example of a metasearch process that a user carries out via MetaLib or a similar product.[2]

A student is interested in the works of Henrik Ibsen. Since the student knows that Ibsen is Norwegian, she submits a search query in several Norwegian resources that she knows about, such as the catalog of the National Library of Norway, the catalog of the University of Oslo, and several archives maintained by the National Library of Norway-the television, radio, and newspaper archives. The student submits the query ***author* = *Henrik Ibsen*** to all these information resources. She then receives the results. If they are displayed by resource, she can easily pick out the results that seem most relevant. Let's say that one result from the television archive is a program about the play *Peer Gynt*, written by Ibsen. Looking at this record, the student decides that she can focus solely on the work *Peer Gynt* rather than all of Ibsen's works. She then uses additional functions of the system to submit a second query, ***title* = *Peer Gynt***, to the same information resources. This time she receives different results, including the *Peer Gynt Suite*, composed by Edvard Grieg - a result from the radio archive that she did not obtain earlier. However, the Ibsen play *A Doll's House,* from the catalog of the University of Oslo, did not come back this time, although it was on the previous result list.

Let's take the process one step farther and ask another question: How did the student know of the resources relevant to her research? Of course, she could have been knowledgeable in this area and aware of pertinent resources. If she was just starting out, however, perhaps she concentrated on the default resources that her library has set, on the basis of her group affiliation, as a component of its gateway. Alternatively, she might have requested resources relevant to her subject, a specific geographic region, a certain type of material, and so on, thus creating a personal searching scope maintained by the system and available for reuse. In MetaLib, such functionality is provided through the Information Gateway component.

One-Stop Shopping

Most researchers today deal with content residing in a wide range of materials. For example, our student might want to access materials such as the script of the play in book form or PDF file, literary analyses of the play, various recordings of the suite, the score of the suite, or a video or poster of a specific performance. The immediate search result is typically a bibliographic record or other form of metadata describing the actual material. From the end user's perspective, the bibliographic records serve only as a means of obtaining the material itself. Users do not want to be bothered with technical issues such as the format of the material they seek and the software that they need to access it - the library OPAC, Adobe® Acrobat® Reader®, Microsoft® Word or PowerPoint®, MP3, the MrSid viewer, or any other software that handles specific types of files.

To provide users with convenient access to materials contained in a range of resources, multiple software products need to be integrated, and they should offer a seamless interface to users. The first type of information typically presented to users as a search result is a description of the material - the metadata - such as a bibliographic record representing a video. Ideally, the user should see the material on her screen - in this case, a video - without having to concern herself about how to find the actual material and how to view it.[3]

The link from the bibliographic record to the actual material can be direct, an explicit URL embedded in the metadata, as in the MARC 856 field of a bibliographic record in library catalogs. However, in many instances, the system must perform calculations to create the link - for example, when the bibliographic record resides in one information repository, such as an abstracting and indexing database, but the actual material resides elsewhere, such as in an e-Journal repository or the library's printed collection. The user expects to reach the actual material nevertheless. A library can make this possible by configuring a context-sensitive linking server, such as the Ex Libris SFX server (Van de Sompel & Beit-Arie, 2001), that links the user to the actual material as a part of a set of extended services and onward navigation options. Such links include the appropriate copy of an article, the holdings in the user's library OPAC or any other relevant OPAC, the institution's document delivery service, citation information, a periodical directory, Internet searches, and information about the book in Internet bookstores or content-based services such as those offered by Syndetics. The software determines the list of links on the basis of the information in the specific bibliographic record and the institution's subscriptions and policies as predefined by the librarians.

Resource Discovery And Information Discovery

The process of finding relevant materials for research falls, therefore, into two stages. First is the resource discovery phase, when the user locates the resources most relevant to the specific search. Next comes the information discovery phase, when the search is executed in the various information resources and the results are retrieved. Institutions strive to provide their members - students, staff, and researchers - with high quality resources that offer information of real value. It is up to the librarians to determine what constitutes the institution's collections, both physical and virtual, and set the collections' boundaries. Every member of the institution should be able to define a personal scope that derives from the institution's scope.

Once the user sets the scope of the search and submits a query, the information discovery phase begins. The metasearch system delivers the query to the selected information resources and returns the results to the user. The process requires that the system 'understands' the expectations of the resources regarding the form of the query, on the one hand, and the nature of the results, on the other. It is up to the system to convert the unified query and adapt it to the requirements of each searched resource, deliver the query in the form appropriate to each resource, receive the results, and manipulate them so that they comply with the system's unified format.

RESOURCE METADATA

The first question, therefore, is which resources are available and which of those are appropriate for the institution. No software can replace librarians when it comes to an understanding of the scholarly information arena; only they can select the resources that are appropriate and affordable for their institution. However, the selection of a resource is just the first step. Information about the resource, *resource metadata*, is necessary as well. The metasearch software needs to obtain descriptive metadata about the resource, such as its coverage and the types of materials that it offers, and makes it available to end users so that they can make a knowledgeable decision about the relevance of the resource to their needs. Furthermore, the system needs technical metadata regarding its impending interaction with the resource.

Resource metadata can be made available in several ways:

1. Resources can offer their metadata to any metasearch system that attempts to access them for the purpose of information retrieval.

2. A central repository can offer resource metadata to any metasearch system.

3. Metasearch systems can maintain their own repository of resource metadata.

The first method - that a resource describes itself when relevant - seems the best. Resources provide the most accurate information about themselves, information that other repositories need not replicate. As a matter of fact, the Z39.50 Explain function was based on this premise. The idea was that when external software needed to access an information resource, the software would extract the details of the impending interaction from the resource on the fly and use the information to formulate the exact steps of the interaction. Apparently, few vendors implemented the Z39.50 Explain function, and those who did implemented it in a variety of forms. The Semantic Web approach takes the idea one step farther. With this approach, a typical metasearch process involves an interaction between agents that exchange requests and information to construct the final product, which is the information requested by the end user. This is the vision, but today's Web does not allow for such interaction between agents, and, therefore, an automated interaction between the metasearch system and a resource's own search engine cannot be achieved at the present time (Sadeh & Walker, 2003).

The second method - building and maintaining a central repository - is under discussion by the new NISO metasearch committee, MetaSearch Initiative, which was formed in early 2003. Maintaining a central repository would assure the availability of resource metadata but would pose new challenges. First, a decision would need to be made about which kinds of resources such a repository would store. Then a format for the resource metadata would need to be specified, as well as protocols dictating the manner in which resource metadata find their way to and from the repository. Finally, a decision would have to be made about who is responsible for storing information in the repository and keeping it updated - the repository, by means of harvesting programs, or the resource itself. Another undertaking similar to that of the NISO committee is the Information Environment (IE) Service Registry pilot project, driven by MIMAS, in the UK, in collaboration with UKOLN and the University of Liverpool. The purpose of the project is to provide a registry of IE collections and services and examine the feasibility of such a registry in terms of discovery, access, maintenance, sustainability, ownership, and scalability. The information science community is watching these initiatives with interest to see whether such repositories become comprehensive and robust enough to provide services as necessary.

The third method is one that various current metasearch products have already implemented. Each such product holds the metadata, both descriptive and technical, of all the resources that it can access. Products differ in the amount of descriptive metadata that they release to the end user and the way in which they display it. They also differ in the degree to which they implement the search interaction and hence vary in the amount of technical metadata that they store.[4]The method whereby each metasearch system maintains information about the resources has many drawbacks. The most obvious one is that every vendor of a metasearch system has to configure and maintain the resource metadata. Handling such a repository requires considerable effort and therefore depends on the capabilities of the individual vendor.

MetaLib, like other products, provides a repository that includes the metadata of all the resources that it can access. However, the metadata are not maintained as part of the software but stored in the MetaLib Knowledge Base, a repository of resource data and rules. The software itself does not include any information that relies on specific resources: it extracts the information from the Knowledge Base. This information enables the user to select the resources and the MetaLib Information Gateway to perform the actual search and retrieval. If, in the future, one of the first two options regarding the origin of the resource metadata materializes, MetaLib will only need to extract the required metadata from another repository.

The Metalib Knowledge Base

The MetaLib Knowledge Base is a proprietary repository provided to institutions along with the MetaLib software. The Knowledge Base holds two types of metadata about resources:

- Descriptive metadata, such as the resource's name, coverage, language, data types, and publisher. The user sees this information and, with it, can make a sensible selection of resources. It is the same information that enables the system to create resource lists based on the user's specifications and display them in a comprehensive way. In short, this information serves the resource discovery phase described earlier.
- Technical metadata, such as the type of protocol that the resource supports, the cataloging format it uses, and the physical and logical structure of the records that it retrieves. We can describe this information as rules that define the flow, interface, and manner of searching and that the software uses for searching, retrieving the results, and manipulating them - that is, for the information discovery phase.
- The resource metadata in the MetaLib Knowledge Base can be divided into global metadata and local metadata:
- Global metadata are that part of the resource metadata that is universal and does not depend on the implementation of MetaLib at a specific institution. These metadata include the name of the resource owner, the coverage, and the interfacing rules.
- Local metadata are institution-specific; they relate to the way in which the resource is used in the institution's environment and presented to the institution's members. Such metadata include elements of authentication vis-à-vis the provider of the resource, the authorization rules that apply to it within the institution, and the categorization information that the institution uses to enable the software to offer the resource in specific contexts. For instance,

one institution might categorize a certain resource under *Medicine,* whereas an institution with a different orientation might categorize it under *Social Studies*.

Ex Libris maintains a master Knowledge Base, which is copied to every MetaLib installation. Automated routines ensure that the Knowledge Base at each installation is updated as necessary. Institutions localize the relevant metadata and add configurations to local resources.

Searching And Retrieving

The process of searching and retrieving in a heterogeneous environment is far from trivial. Each resource has its own expectations regarding the form and manner in which it receives queries; even if the resource supports a standard interface, such as the Z39.50 protocol, the metasearch system needs to make further adjustments so that the resource's engine will interpret the query correctly.

The types of information that the Knowledge Base maintains to enable the system to search include the following examples:

- Access mode: What kind of interfacing protocol does the resource employ? Is it a structured, documented interface, such as Z39.50, the PubMed Entrez protocol, or a proprietary XML gateway? Or is it an unstructured HTTP protocol that dictates the use of HTML parsing techniques to access the resource?
- Password control: How does the user access a specific, licensed resource? Are a user ID and password required, which the metasearch system delivers when the connection is established? Should the software redirect the query via a proxy to grant the user access?
- URL creation: If a URL needs to be formulated to hold the specific query, what should the structure of the URL be?

- Character conversion: What character set does the system use at the resource end? Does the character set comply with that of the end user?

- Query optimization: How should the query be structured?

- What is the exact syntax that the resource's system expects?

- How should fields be mapped to the fields of the resource; for example, to which field should the system map the "author" field selected by the user for a specific query?

- How does the system expect to receive an author's name? Should it be <last name><,><first name>; <last name>< ><first initial>; or in some other format?

Normalization: What should the system do when the search engine at the resource end does not support a specific type of search? For instance, what rules should be applied if the user looks for a specific subject but a certain resource does not support a search by subject?

Once the information is there, the metasearch system can indeed adapt a single, unified query to the requirements of the specific resource, as in the following example.

The user submits a query for ***title = dreams*** *and* ***author = Schredl, Michael*** in the following resources:

- Library of Congress (Z39.50 access to Endeavor's Voyager ILS)

- NLM PubMed (the Entrez HTTP protocol)

- HighWire Press® (HTML parsing)

- Ovid MEDLINE® (Z39.50 access via the SilverPlatter ERL platform)

- University of East Anglia (XML access to the Ex Libris ALEPH ILS)

Even when looking at one brick of the process structure - the query syntax - we can clearly see the differences between the resources:

- The Library of Congress expects this query string:

*1=**Schredl, Michael** AND 4=**dreams***

- PubMed expects this query string:

*term=**dreams**+AND+**Schredl+M***

- HighWire expects to see the encoded form of the following URL:

*author1=**Schredl,+Michael**&author2=&title=**dreams***

- Ovid's MEDLINE via ERL, although accessed by the same protocol (Z39.50), expects this query string:

*1003=**Schredl-M*** AND 4=**dreams***

(Note the phrasing of the author's name.)

- The ALEPH system at UEA expects the following encoded request:

*wau=**(Schredl, Michael)** AND wti=**(dreams)***

Presentation Of Search Results

Up to now we have discussed only the flow from the user to the resource. However, now that the query has been processed, the metasearch system needs to get back to the user with search results. Typically the interaction between the metasearch system and the resource consists of two phases. The first occurs after the search has been invoked: the resource returns the number of hits and some kind of reference to the result set. This phase is important because it gives the user some information about the search and enables the user to refine the query before browsing through the results. For instance, if a user sees that there are thousands of hits, she can modify the query to be more specific and thus reduce the number of results. The second phase consists of retrieval: The metasearch system retrieves the number of hits along with the first few records for each resource. This information is shown to the user instantly, even though the query might result in hundreds or thousands of hits. Some systems, including MetaLib, allow for further retrieval upon request.

Why do the systems provide such limited retrieval initially? First, retrieval depends on the use of networks, which are still not as rapid as one would like. Retrieving hundreds or thousands of records over a network is an extremely time-consuming process, and users are not likely to wait until it is completed. Second, people have difficulty handling immense result sets; after seeing the number of hits for each resource, users are likely to refine their query to obtain fewer hits. Once retrieved from the resource, each result is converted to a unified format before the user sees it. The rules that define the manipulation of the retrieved data are part of the resource metadata, which, in MetaLib, is stored in the Knowledge Base. These rules include information about the logical format, the cataloging format, the script, and the structure of certain fields, such as the citation field. For further processing to take place, the metasearch system must be able to apply these rules and convert all retrieved records, regardless of their origin.

Such additional processing can include the unified display of the records to end users; the merging of result lists from heterogeneous resources into one list; the comparison of records to eliminate duplicates; the creation of an OpenURL to allow context-sensitive reference linking; and the saving of records in whatever format is required. Consequently, functionality that might have been missing from the native interface of the resource, such as the provision of an OpenURL, is added to the same set of records by the metasearch system. However, the display of result lists is not as straightforward as might be expected. Users are well acquainted with Web search engines and therefore have solid expectations regarding the display. They would like their results ranked, merged into one list, and filtered for a selected resource. Furthermore, they would like to be able to sort results by various attributes, such as title, author, and date.

Given that only the first results are retrieved from the various resources, these expectations are not so easily satisfied. When the result sets are small, all records are in the system's cache memory and so the metasearch system can offer the expected functionality in a comprehensive manner. However, the larger the number of hits, the greater the value of merging, sorting, de-duplication, and ranking - and the more difficult these features are to provide. Consider, for instance, the merging of the lists. How should it be done? The number of hits may vary considerably from resource to resource. Would it be appropriate to merge the two hits received from one resource with the dozens or hundreds of hits received from another resource? And if so, in which order? Every resource returns results in a different sorting order - by date (ascending or descending), title, relevance, or another attribute of which the users are not necessarily aware. Because only the first records are retrieved, the issue of merging the results needs careful consideration.

Other issues are the sorting capability and relevance ranking that users expect to find when looking at results, even when the resource itself does not support such functionality. Does it make sense to rank and sort only those results that have been retrieved? Let's say that the metasearch system applies certain relevance-ranking algorithms to all retrieved records and sequences them accordingly in the display to the user. This display can be rather misleading, because the 'best' hits are not necessarily those that were retrieved first. It could well be that if the user asks for more hits better results will be retrieved. A similar problem applies to sorting: even though a system might enable the user to sort the records according to various parameters, this sorting would apply only to the set already retrieved.

MetaLib handles these issues by always allowing the user to see the results for each resource. If the resource supports sorting, the user can request that the result list be sorted. Then MetaLib submits the search to this resource again, asking that the entire set of results be arranged in the order requested by the user. Hence, the user indeed receives the first records of the whole set. MetaLib also enables users to explicitly request and obtain a merged set at any point. Such a set is already de-duplicated and sortable. Institutions are likely to limit the number of records that can be merged, to avoid lengthy waiting periods caused by the retrieval of large result sets.

Local Repositories And Local Indexes

End-users may wonder why other searching systems, primarily the Web search engines, are able to provide them with large sets that are merged and ranked. The reason is that these systems use a different type of technology to provide the users with search results. Metasearch systems are based on 'just-in-time' processing. The system does not maintain any indexes of its information landscape locally; only when the information is required does the system access the various resources to obtain the results. The approach of Web search engines is based on 'just-in-case' technology. Huge efforts are invested in preparing the information prior to users' requests so that when the information is needed, it is obtained immediately. Google™, for example, holds indexes for the entire World Wide Web, including not only pointers to sites but also information that enables the search engine to evaluate the relevance ranking of a site. When the user searches with Google, only the indexes are scanned - and the information that Google initially displays on the screen is not from the sites themselves but from this vast repository of indexes. The search engine provides the actual access to a certain Web location only when the user selects it from the list. Needless to say, huge computing power and disk space along with sophisticated technologies for harvesting, evaluating, and maintaining the information are necessary for such powerful tools.

The use of local repositories of indexes in the library environment started some time ago. As opposed to union catalogs, which actually replicate the information that is located in local catalogs, repositories such as MetaIndex from Ex Libris hold only the indexes to the bibliographic materials that are kept in the resources. An example is the MetaIndex implementation at the Cooperative Library Network Berlin-Brandenburg (KOBV), which preceded the metasearch systems a few years back: At KOBV, MetaIndex enables each of the consortium members to maintain its library system and cataloging conventions while the consortium provides a single search interface for end users. MetaIndex has now become a resource available to MetaLib at KOBV, along with other resources. No doubt that a local repository of indexes has many advantages. Information that is gathered and processed prior to queries can be organized, evaluated, and de-duplicated and therefore can be accessible to end-users in a rapid and comprehensive manner. However, maintaining such a repository has a major drawback: the repository is another system, with hardware and software, to create and maintain, and personnel must be available to take care of it.

Considering libraries' budget constraints and limitations in the technical expertise available to them, a combination of just-in-case and just-in-time approaches would be optimal for metasearch systems. Local repositories would be useful in the following cases:

- When no searching mechanism exists at the resource end. This situation is typical of various types of local repositories, such as those that hold research papers written by institution members or spreadsheets relevant to institutional activities; but it could obviously apply to any other data that have not yet been made available to the public.
- When the information is scattered. A local repository may be worthwhile if several resources that are mutually compliant form a single resource of value to the institution. For example, a worldwide organization that has dozens of branches, each of which holds regionally relevant information, wants to provide a simultaneous search capability that will cover all the local information. Creating an index such as MetaIndex would be preferable to requiring users to search all the repositories simultaneously.

- When the interface is not reliable. Some institutions want to provide access to resources that are not always online or do not offer reliable networking for accessing them. In such cases, an institution might be better off harvesting the information and keeping it as a local repository.

- When preprocessing is important. Preprocessing tasks such as relevance ranking and the elimination of duplicate records can be of value for some institutions. However, a component like MetaIndex can provide a solution only if the search scope is defined and limited. For instance, at KOBV, the consortium catalogs represent a limited search scope; as a result, the mathematics department of the consortium was able to develop a sophisticated de-duplication algorithm that permitted the construction of a comprehensive MetaIndex component.

MetaIndex from Ex Libris is created through the harvesting of information from other repositories. One of the harvesting mechanisms is the Open Archives Initiative Protocol for Metadata Harvesting (OAI-PMH). The use of such harvesting protocols can facilitate the gathering of data and is applicable to a wide range of resources that are now becoming OAI compliant. Furthermore, MetaIndex itself can become OAI compliant, thus serving as both a resource for MetaLib and an OAI-compliant resource that enables other systems to harvest the data from it.

SUMMARY

The promise of a truly integrated environment in a heterogeneous world may not yet be a reality, but with the active involvement of all the stakeholders, significant progress has been made. Just a few years back, metasearch systems seemed like a dream; today they are already a building block in the information resource environment serving the academic and research community.

REFERENCES

1. Sadeh, T. & J. Walker.: "Library portals: toward the semantic Web". *New Library World* 104(2003)1/2, 11-19.
2. Van de Sompel, H. & O. Beit-Arie.: "Open Linking in the Scholarly Information Environment Using the OpenURL Framework". *D-Lib Magazine*, 7(2001) 3.
3. http://www.dlib.org/dlib/march01/vandesompel/03vandesompel.html

NOTES

[1] The term federated searching is used by some to describe a process in which indexes are 'pregenerated'. We refer to this concept as 'just-in-case' processing, as explained later in this paper.

[2] We provide this example only to illustrate the process; references to specific resources are not necessarily accurate.

[3] The issue of copyrights is not discussed in this paper. In this context, we assume that the system that offers the material handles the copyright issues.

[4] For instance, some products offer unified searching, but once the user requests the result record, the software links the user to the record in the resource's native interface. Such products do not need to maintain all the technical metadata that is required for manipulating the retrieved record and converting it to a unified format.

Chapter 12

Going beyond Citations: SERUM — a new Tool Provided by a Network of Libraries

Juan Gorraiz, Christian Gumpenberger

University of Vienna, Library and Archive Services, Bibliometrics

Department, Boltzmanngasse 5, A-1090 Vienna,

ABSTRACT

Citation metrics are a well-established means to assess the impact of scholarly output. With the growing availability of e-journals, usage metrics have become an interesting alternative to citation metrics; they allow for viewing scholarly communication from the user's perspective. Usage metrics offer several advantages that have the potential of enhancing existing evaluation criteria for scholarly journals. This paper suggests an approach to providing global usage metrics which is supported by libraries. The goal is to provide an analytical tool called Standardized Electronic Resource Usage Metrics (SERUM) which is comparable to the Journal Citation Reports (JCR), but which makes use of download data instead of citation data. Global download data would be obtained from the publishers, assuming they are willing to contribute in order to benefit from newly established evaluation criteria for periodicals beyond the Journal Impact Factor (JIF) and consequential strengthening of their products. An international network of libraries with a sound disciplinary coverage will be established in order to obtain, manage and check the authenticity of the data delivered by the publishers. The network will act as a clearing centre operated by independent information specialists to guarantee data integrity as well as data curation according to a standardised format. Furthermore, the network's internationally distributed members should also track and manage local usage data, reflect local trends, and relate these to the global publishers' data. In addition, a regularly updated version with specific basic usage metrics and journal rankings will be offered.

Keywords: SERUM, usage metrics, scholarly journals

INTRODUCTION: CITATION AND USAGE METRICS — AN OVERVIEW

In 1927 the Gross brothers introduced the system whereby the total number of citations (excluding journal self-citations) are used to identify the most relevant journals in a research area (Gross and Gross, 1927). For the first time citations were used as a collection management instrument. Before that time, library acquisitions departments relied mainly on usage data. But tracking physical usage was a cumbersome and difficult task. The data obtained in this way were usually combined with information obtained from user surveys, although these surveys were always perceived as time-consuming as well as annoying for the library users. Document demand patterns apparent from document delivery services were also used for decision making in academic librarianship; the relationship between the most requested and the most cited journals was studied e.g. by Schloegl & Gorraiz (Schoelgl and Gorraiz, 2006). The introduction of the Science Citation Index by Garfield in 1964 laid the foundation for a whole new discipline called scientometrics. In 1972 Garfield published his article 'Citation analysis as a tool in journal evaluation' in Science and it attracted a great deal of attention from journal editors (Garfield, 1972). Three years later Garfield launched the first edition of the 'Journal Citation Reports' offering global journal rankings. This tool, initially called 'a bibliometric analysis of science journals in the ISI database', introduced the following citation metrics:

1. Total cites: as a measure for the citation quantity
2. Impact factor: as a measure for the average citation frequency
3. Immediacy index: as a measure for the citation speed
4. Cited/citing half-life: as a measure for the ageing characteristics of a subject field.

Meanwhile, citation metrics have become well established (especially in Science, Technology and Medicine) and their benefits are clearly promoted by Thomson Reuters' JCR for the different target groups:

- Librarians can manage and maintain journal collections and budget for subscriptions (...).
- Publishers can monitor their competitors, identify new publishing opportunities, and make decisions regarding current publications.
- Editors can assess the effectiveness of editorial policies and objectives and track the standing of their journals.
- Authors can identify journals in which to publish, confirm the status of journals in which they have published, and identify journals relevant to their research.
- Information analysts (science policy makers) can track bibliometric trends, study the sociology of scholarly and technical publications, and analyse citation patterns within and between disciplines.

The importance and necessity of analytical citation tools like JCR is supported by the launch of alternative products like SCImago Journal & Country Rank and CWTS Journal Indicators, both attempting to compensate for the shortcomings of JCR or to include new features.

The growing availability of e-journals due to the advent of the internet and their increasing acceptance resulted in a rapid change in the user preference especially between 2001 and 2006, as reported in several studies (Kraemer,

2006; Schloegl and Gorraiz, 2010). As a result, usage metrics were reintroduced as an interesting alternative to citation metrics. Compared to the printed era, data collection is now much easier and faster, and usage metrics allow for viewing scholarly communication from the user's perspective.

The correlation between citation and usage data is highly dependent on the discipline's publication output and has been well documented in several studies (Bollen et al., 2005;Moed, 2005; Brody et al., 2006; Wan et al., 2008). Usage metrics can therefore be regarded as complementary to citation metrics; they reflect usage in a much broader scope (Duy and Vaughan, 2006) and have caused the emergence of the field of bibliometric research (Bollen and Van de Sompel, 2008; Schloegl and Gorraiz, 2009).

This paper suggests an approach to providing global usage metrics which is supported by libraries. The goal is to provide an analytical tool called Standardized Electronic Resource Usage Metrics (SERUM) which is comparable to the Journal Citation Reports (JCR), but which makes use of download data instead of citation data.

Usage versus Citations

Citations metrics have a few well-known disadvantages. The most important ones are:

1. Citations are highly controversial. There are different reasons why authors cite others (objectivity vs. subjectivity; the problem of self-citations, citation networks, etc.).

2. Citations are only an indirect measure for impact, and no indication for quality.

3. Citations are always delayed. In some disciplines, half-lives cause citations to become significant only after long time spans.

4. Citations are simply insufficient in some fields. There are surprisingly low mean citation rates for the vast majority of published scholarly literature.

In comparison usage metrics overcome some of the disadvantages mentioned before:

1. Downloads are generally accepted as a proxy for usage.

2. Downloads are a direct measure and usage data can be recorded automatically.

3. Usage data can be accessed without delay.

4. Downloads are more suitable than citations in less publication-intensive fields.

There are of course known issues to be sorted out with usage data:

1. The availability of global usage data is restricted.

2. There is a risk of inflation by manual or automatic methods.

3. There are different access channels to scholarly resources (e.g. publisher sites vs. subject repositories vs. institutional repositories).

SERUM — Necessity, Scope and Requirements

SERUM (Standardized Electronic Resource Usage Metrics), is an initiative which was suggested in response to the lack of globally available usage metrics for scholarly communication channels (primarily e-journals). Usage metrics definitely add a new dimension to existing quality criteria for scholarly journals, which are currently — apart from citation metrics — restricted to peer review, timeliness, editorial excellence, language and bibliographic information. As scholarly journals are still perceived as the most important scholarly communication channel, SERUM's initial focus is on the journal level.

The idea of SERUM is to combine the following three major features:

1. providing access to consolidated global usage data of e-journals;

2. establishing an international network of libraries that relates local usage trends to global usage data provided by the publishing houses;

3. offering services like ‘usable’ usage metrics and journal rankings (Figure 1).

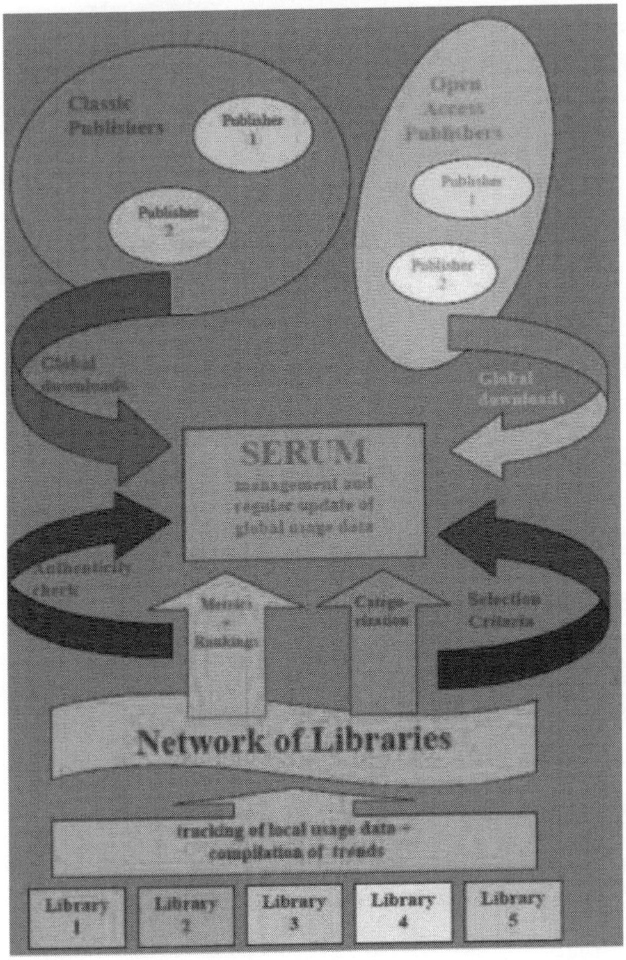

Figure. 1 Schematic overview of SERUM.

ad 1. Global Data Obtained from the Major Publishers

In order to pave the way for SERUM it will be necessary to win over the publishers. Experience has shown that the big traditional publishers are more hesitant to cooperate than open access publishers. But both categories of publishers have a need to gather information about the usage of their products and track changes on the demand side of the market. Therefore publishers should be highly interested in the project since:

- they would benefit from newly established evaluation criteria for periodicals beyond the JIF, and
- a well-used initiative like SERUM would mean consequential strengthening of their products.

How is SERUM to work? Each publisher is invited to deliver usage data from as many titles per subject field as possible. The focus should of course be on top titles in terms of usage. From these titles, a network of libraries (see below) makes a selection of titles to be used in SERUM and for their categorisation using the most appropriate classification scheme.

The following data would be required (all at journal level) from the publishers:

- total number of downloadable items per journal
- number of downloadable items disaggregated (by document types) per journal
- total download counts (full-text article requests — FTAs) for the current year (independent from the publication years of the journal)
- download counts (full-text article requests — FTAs) for the current year with explicit listing of downloads per publication year of the journal for the last 5 years
- download counts at journal level disaggregated by following document types: articles, review articles, proceedings papers, letters, notes
- percentage of 'non-downloaded' items
- distinction of downloads according to their origin (grouping by identical IP addresses without disclosing them) ↔ total view and geographical grouping (according to geographical distribution of the SERUM library network)
- The following data could be provided optionally (all at journal level):

- download counts (full-text article requests — FTAs) disaggregated by treatment or content (e.g., theoretical, methodological and experimental)
- information about the distribution of downloads (how many articles (%) accumulate 75% of the total number of downloads, quantiles or quartiles)
- total download counts (full-text article requests — FTAs) of the retrospective 1–5 years
- distinction in regard to merits of the 'downloads': academic vs. non-academic, etc.

The data obtained would be consolidated and managed by the network of libraries and made accessible via a cooperative website offering further usage metrics and services based on these data (see below). This website will be primarily designed for journals analysis, but can of course be extended to serials and e-books later on. The outcome is a new instrument for the evaluation of electronic resources going beyond citations.

ad 2. Implementation of a Network of Academic Libraries

Single publishers are obviously not in the right position to deliver and manage overall global usage data obtained from multiple publishing houses. Moreover, suggested metrics and services should be provided by an independent, non-biased institution, as self-beneficial data manipulation by single publishers cannot be excluded. Therefore an international network should be established consisting of academic libraries with a sound disciplinary coverage and a significant geographical representation. This network would be responsible for fulfilling the following tasks:

- implementation of the selection procedure for considered journal titles
- categorisation of journals
- management and regular update of global usage data either harvested from or delivered by publishers
- introduction and promotion of usable metrics

- provision of journal rankings
- tracking of local usage data
- authenticity check of global usage data based on local trends.

ad 3. The Metrics in SERUM

Several usage indicators have been suggested in recent years. Most suggestions are based on the classical citation indicators from the Journal Citation Reports (JCR), using download data (usually full-text article requests) instead of citations. The corresponding usage metrics are usage impact factor (UIF) (Rowlands and Nicholas, 2007; Bollen and Van de Sompel, 2008), usage immediacy index (Rowlands and Nicholas, 2007), download immediacy index (Wan et al., 2008), and usage half-life (Rowlands and Nicholas, 2007).

SERUM intends to develop a novel approach to usage metrics. Because the majority of downloads are made in the current and preceding year (Schloegl and Gorraiz, 2010), the classical immediacy index and the usage impact factor (UIF) seem to render less relevant results. Instead we suggest:

- The journal usage factor (JUF) takes into account the reference year and the preceding two years. The JUF is defined as the number of downloads from journal items published in the current year and the previous two years divided by the number of items published in these three years. A three-year time window assures that a very significant number of downloads is covered in most cases.
- The download immediacy (DI) will be measured as the percentage of articles downloaded in the current year in relation to the total amount of downloads.
- The download half-life (DHL) is defined by the number of publication years from the current year that account for 50% of the downloads of the journal.

Further characteristics of the anticipated metrics in SERUM include:

1. Only downloads (full-text article requests — FTAs) will be considered, no visits or hits.

2. SERUM will operate at journal level and not at article level. SERUM will be designed as an instrument to reflect the usage characteristics of a journal as an entity. The four basic indicators are DT (total number of downloads), JUF, DI and DHL.

3. SERUM aims to disaggregate on the basis of document type (articles, review articles, proceedings papers, letters, notes and others); in other words, calculating JUF, DI, DHL for each document type.

4. If possible, SERUM also aims to disaggregate on the basis of treatment or content (e.g., theoretical, methodological and experimental) as well as in regard to the merits of the downloads (academic vs. non-academic).

5. SERUM's usage metrics will work at synchronic level (= downloads are tracked from one fixed year of documents issued in two or more publication years) as well as at diachronic level (= downloads are tracked from two or more years of documents issued in a fixed publication year). Also, all indicators will be calculated for the current and the previous two years, and the results will be compared.

6. According to the skewness distribution hypothesis of downloads at article level in a given journal, we are also interested in other distribution parameters, like the percentage of 'non-downloads', quantiles or quartiles.

7. All indicators will be calculated as weighted indicators on the basis of unique URLs, (the origin of downloads) in order to detect data manipulation.

8. Time lines and graphs will be provided for the most relevant indicators.

9. Increase/decrease rates will be analysed.

10. Once SERUM has been operational for a period of time, the metrics will be corrected in order to reflect usage practices and traditions in the different fields and disciplines. After establishing metrics for e-journals, corresponding metrics will be provided for e-books as the next logical step.

SERUM in the Context of Initiatives to Date

With the advent of the internet and the increased usage of e-journals, users of scholarly information migrated to the digital environment. As a consequence, 'publishers have inherited all the knowledge of the user the librarian once had' (Jamali et al., 2005). Libraries now depend on the usage data provided by the publishers.

Gathering local usage statistics is a cumbersome activity due to the technical requirements involved. Nevertheless the idea of comparing locally collected usage data to vendor-provided statistics was born seven years ago (Duy and Vaughan, 2003) and picked up by Coombs (2005). SERUM aims to revive this idea and build on the experience gained from previous studies as well as benefit from ongoing harmonisation and improved provision of vendor

statistics as a result of COUNTER and SUSHI. Like MESUR, SERUM also aims to obtain usage data from the publishers in order to derive usage metrics from these. However, other than MESUR it is neither the objective of SERUM to create a complex semantic model of the scholarly communication process nor to produce science maps (Bollen et al., 2009) on the basis of the established reference data set. The maps and ranking services

available from ' MESUR: science maps and rankings from large-scale usage data' are highly complex and only insightful for a minority of experts.

The unique aspects of SERUM will therefore be:

- relating locally tracked usage data to global usage statistics provided by publishers (authenticity check)

- strengthening the role of academic libraries by implementing the network with the suggested scope of duties
- introducing new aspects in the evaluation of journals for all mentioned target groups (librarians, publishers, editors, authors, information analysts)
- separating usage from citations
- focusing on simplicity and usability.

SUGGESTED APPROACH FOR SERUM PILOT

Before all else, it is crucial to win over a few publishers. Large traditional publishers have already been approached on the subject and have either not responded or have refused their cooperation. As a consequence it was decided to address open access publishers instead to get the process started. Two large and one small OA publishers have responded positively and are willing to contribute to the SERUM project. The next step will be to share and discuss the data requirements and the feasibility of delivery.

Also, the idea of SERUM will be promoted amongst academic libraries (e.g. at conferences like LIBER) in order to identify potential partners for the network of libraries. Apart from the University Library of the Vienna University at least two further partners would be required for the pilot. Depending on the international reaction the requirements specification will be elaborated and a project plan designed.

CONCLUSIONS

Quite a few metrics initiatives have focused on usage data. However, so far none of them combined all the major features of SERUM, i.e., access to consolidated global usage data of e-journals; implementing an international network of libraries that relates local usage trends to global usage data provided by the publishing houses; offering services like 'usable' usage metrics and journal rankings.

Download statistics are not only relevant and necessary at the article level. The main scholarly communication channels themselves need to be measured, analysed and evaluated as entities (focus on journals, but also on monographs, etc.). Obviously, the prestige and eminence of journals is of major interest to many stakeholders in the process (librarians, publishers, editors, authors and information analysts).

Citations alone show only a part of the whole picture and are insufficient especially in the social sciences and in the arts & humanities. This is demonstrated by ERIH, (European Reference Index for the Humanities), a unique peer-review project in this discipline that aims to compensate for the unsatisfactory coverage of European humanities research in existing citation indices (AHCI, SSCI). However, this peer-review project is subject to continual revision and development, and support is highly expensive. Downloads would offer a more feasible alternative, but usable download metrics and services for a broad audience are not yet available (MESUR has a different scope).

At this point in time SERUM is only a 'blue sky project'. Nevertheless its benefits are obvious and easy to summarise:

- simplicity
- comprehensibility
- usability
- fast and central availability.

REFERENCES

1. Bollen, J., H. Van de Sompel, J.A. Smith and R. Luce (2005): 'Toward alternative metrics of journal impact: a comparison of download and citation data', *Information Processing and Management* 41, 1419–1440; http://public.lanl.gov/herbertv/papers/ipm05jb-final.pdf (accessed 1 July 2010).

2. Bollen, J. and H. Van de Sompel (2008): 'Usage impact factor: the effects of sample characteristics on usage-based impact metrics', *Journal of the American Society for Information Science and Technology*, 59(1), 136–149.

3. Bollen, J., H. Van de Sompel, A. Hagberg, L. Bettencourt, R. Chute, M.A. Rodriguezn and L. Balakireva (2009): 'Clickstream data yields high-resolution maps of science', *PLoS One* 4(3), e4803, doi:10.1371/journal.pone.0004803.

4. Brody, T., S. Harnad and L. Carr (2006): 'Earlier web usage statistics as predictors of later citation impact', *Journal of the American Society for Information Science and Technology*, 57(8), 1060–1072.

5. Coombs, K.A. (2005): 'Lessons learned form analyzing library database usage data', *Library Hi Tech*, 23(4), 598–609.

6. Duy, J. and L. Vaughan (2003): 'Usage data for electronic resources: a comparison between locally collected and vendor-provided statistics', *The Journal of Academic Librarianship*, 29(1), 16–22.

7. Duy, J. and L. Vaughan (2006): 'Can electronic journal usage data replace citation data as a measure of journal use? An empirical examination', *The Journal of Academic Librarianship*, 32(5), 512–517.

8. Garfield, E. (1972): 'Citation analysis as a tool in journal evaluation: Journals can be ranked by frequency and impact of citations for science policy studies', *Science,* 178(4060), 471–479.

9. Gross, P.L.K. and E.M. Gross (1927): 'College libraries and chemical education', *Science,* 66(1713), 385–389.

10. Jamali, H.R., D. Nicholas and P. Huntington (2005): 'The use and users of scholarly e-journals: a review of log analysis studies',

Aslib Proceedings : New Information Perspectives, 57(6), 554–571.

11. Kraemer, A. (2006): 'Ensuring consistent usage statistics, part 2: working with use data for electronic journals', *The Serials Librarian*, 50(1/2), 163–172.

12. Moed, H.F. (2005): 'Statistical relationships between downloads and citations at the level of individual documents within a single journal', *Journal of the American Society for Information Science and Technology*, 56(10), 1088–1097.

13. Rowlands, I. and D. Nicholas (2007): 'The missing link: journal usage metrics', *Aslib Proceedings : New Information Perspectives*, 59(3), 222–228.

14. Schloegl, C. and J. Gorraiz (2006): 'Document delivery as a source for bibliometric analyses: the case of Subito', *Journal of Information Science*, 32(3), 223–237.

15. Schloegl, C. and J. Gorraiz (2009): 'Global usage vs. global citation metrics using Science Direct pharmacology journals', *Proceedings of the International Conference on Scientometrics and Informetrics*, 1, 455–459.

16. Schloegl, C. and J. Gorraiz (2010): 'Comparison of citation and usage indicators: the case of oncology journals', *Scientometrics*, 82(3), 567–580.

17. Wan, J.-K., P.-H. Hua, R. Rousseau and X.-K. Sun (2010): 'The download immediacy index (DII): experiences using the CNKI full-text database', *Scientometrics*, 82(3), 555–566.

Index

S

U

V